TWELVE INVENTIONS
WHICH CHANGED AMERICA

The Influence of Technology on American Culture

Gerhard Falk

Hamilton Books
A member of
The Rowman & Littlefield Publishing Group
Lanham · Boulder · New York · Toronto · Plymouth, UK

Copyright © 2013 by
Hamilton Books
4501 Forbes Boulevard
Suite 200
Lanham, Maryland 20706
Hamilton Books Acquisitions Department (301) 459-3366

10 Thornbury Road
Plymouth PL6 7PP
United Kingdom

All rights reserved
Printed in the United States of America
British Library Cataloging in Publication Information Available

Library of Congress Control Number: 2013930407
ISBN: 978-0-7618-6080-8 (paperback : alk. paper)
eISBN: 978-0-7618-6081-5

⊖™ The paper used in this publication meets the minimum
requirements of American National Standard for Information
Sciences—Permanence of Paper for Printed Library Materials,
ANSI Z39.48-1992

Dedication

This book is dedicated to the government and people of the United States, who let me live.

Contents

Acknowledgments

I acknowledge the help of my son Clifford, without whose help this book could not have been written.

Introduction

The History and Influence of Technology and Science on American Culture

Technology and Science are not only material objects, but also consist of social relations, ideas and the sociological imagination. Culture is the man made environment consisting of material culture, behavioral culture, and ideological culture. Therefore it is evident that changes in material culture promoted by discoveries and inventions have consequences for all three of these aspects of culture. The evidence for this can be seen by looking at the introduction of technological advances into primitive cultures such as the Yir Yoront culture of Australia. That culture literally fell apart after these Stone Age people were given steel axes by Australian "do gooders" in 1915. That innovation replaced stone axes, with devastating consequences for these primitive people.[1]

The view of technology and science at the beginning of the second decade of the 21st century is quite different from that common in the 19th century. Then it was believed that the world would change for the better in an unstoppable cascade of new inventions and discoveries and improvements, as visible in the Crystal Palace exhibition in London in 1851. That optimism has now been replaced with the hope that things won't get worse, as indicated by the atomic disaster at Fukushima, Japan and the ever increasing poverty in the so-called third world.[2]

Then there is demography. In the western world and in Japan, the old are outnumbering the young, leading either to the need of the working population to support more and more unemployed old people or to the need of old people to kept working longer and longer.

In the 21st century it is also evident that the world needs more and more sources of energy as the population gets larger and larger, reaching nearly 7 billion in 2012. No doubt inventors and engineers are already facing the problem of gaining more sources of energy than is now available from steam, gas, oil, and limited atomic power.

Of these, steam was the first means of creating the industrial revolution, which freed men and animals of much physical labor after James Watt (1736-1819) improved the steam engine in the second half of the 18th century by tripling its efficiency.

Thereafter, a literal flood of inventions has benefited mankind and particularly the people of the United States, with many consequences not foreseen.

This book will be of interest to the general reader as well as students in the sociology of work and in the history of technology.

Notes

1. Lauriston Sharp, "Steel Axes for Stone Age Australians," IN: Edward H. Spice, *Human Problems in Technological Change,* (New York: The Russell Sage Foundation, 1952), pp. 69-90.

2. Tobin A. Sparling, *The Great Exhibition: A Question of Taste,* (New Haven: Yale Center for British Art, 1983).

Chapter One

The Steam Engine

I.

James Watt did not invent the steam engine. However, his improvements were the catalyst which brought on the Industrial Revolution in England and America. That means that a third phase of empowerment had arrived in the history of human culture. In the first stage, the longest period of human development, human muscle alone was available. The second stage in the history of empowerment occurred approximately around 9000 B.C., with the domestication of animals, beginning with the dog.

The steam engine was evidently the first device which succeeded in substituting a machine for the power of animals. It was Thomas Savery (1650-1715) who in 1698 made a water pumping device. It was based on a use of steam power, as suggested by the Marquis of Worcester in 1663. Savery's engine was used to lift less than 40 pounds in the tin mines of Cornwall, where a mechanic, Thomas Newcomen (1663-1729), constructed a steam engine with a piston. That engine was used almost entirely for the raising of water or the turning of a water wheel. Thereafter a number of inventors added modifications to the Newcomen engine, until James Watt introduced a separate condenser. After a number of failures, Watt succeeded in 1776 in introducing a working steam engine, with immense consequences for England, America, and the world.[1]

It took many years before the steam engine was put to use in agriculture and in transportation. In 1816, Luke Johnson of Leominster, Massachusetts, patented a machine for drawing a plow. Later in the nineteenth century, a number of inventors experimented with a steam plow all over the western United States. That led to the developments of steam plows, so that eight men could do the work of twenty-four teams at one quarter of the initial investment and one third of the operating expenses. Steam also increased production per acre.[2]

In the middle of the 19th century, ninety different types of steam plows were invented by Americans. Nevertheless, Americans were reluctant to use the new invention, which became widely used in England, mainly because of "the English wizard" John Fowler. His device cost between \$10,000 and \$12,000, which was a huge investment at that time.[3]

In America, several methods of applying steam to plowing were invented, although either these devices were not satisfactory or, as was true of the majority of American farmers, the farmers were not willing to try something new. It wasn't util the 1890's that the steam plow was so far improved that it was finally accepted by American farmers. Then, plows were huge and had the capacity of churning up fifty to seventy acres per day. Yet, the steam engine did not last long, as it was soon replaced by the internal combustion engine.[4]

II.

Although the steam engine had little influence on American farming, its role in American transportation was immense, and continues in a diminished form today. Steam engines became the dominant means of transportation on water and on land. On land, the steam engine became the locomotive, which promoted the development of the railroads in the 19th century. This development may be seen by inspecting the employment statistics in the railroad industry since 1920, when two million people worked for the railroads. Thereafter, rail transportation declined, so that by 1947 that employment had been reduced to 1, 16,000 until, in 2009, only 226,000 people worked for America's railroads.[5]

The first American steam engine was imported from England in 1829. Named "The Stonebridge Lion," as soon as it arrived, it was placed on a railway and driven one mile, From Carbondale to Honesdale, Delaware. That engine had four wheels with a boiler resting directly on the axle. This was two years after Matthias William Baldwin (1795-1866) had begun building stationary steam engines, with the inevitable consequence that Baldwin soon built a locomotive called Old Ironsides. The year was 1832 when that engine first traveled from Philadelphia to Germantown, taking the place of horses when the weather was "fair."[6]

Baldwin continued for some time to produce stationary steam engines, until in 1836 the locomotive business reached more than fifty percent of the Baldwin construction volume. By 1838, there were about 350 steam locomotives in the United States. Of these, Baldwin had constructed 122. There were then twenty-three shops in the United States constructing locomotives. Baldwin was the largest of these. In the 1830's, a number of locomotives were still imported from England, but by then Baldwin was producing about 45 percent of all American locomotives.

The cost of locomotives in the 19th century seems unbelievably low when judged from the 21st century. Between 1834 and 1836, the price of a locomotive ranged between \$6,317 and \$6,400. That price was raised to \$7,000 in 1837, where it remained for some time.[7]

It was in 1837 that the United States suffered a major recession, leading to mass unemployment, wholesale bankruptcies, innumerable foreclosures, and increasing unpaid debt. This led to a sharp decline in the building of locomotives and other industrial products, although by then there were seventy-eight locomotives running on American rails.[8]

The most important outcome of these advances in building stream locomotives was the opening of the American west all the way to California. This became possible when the golden spike was driven into the last rail at Promontory Point in Utah on May 10, 1869. President Abraham Lincoln had signed the Pacific Railroad Act in 1862, so that the railroads thereafter ran through northern territory, causing trade to run east to west instead of north to south. This was a political action which contributed to the civil war victory of the north over the south by 1865. No doubt, the north had a great advantage over the south because the north was already industrialized by 1861. The use of the steam engine as a locomotive made it possible for northern troops to use trains with guns and turrets to shoot at southern troops. The locomotive also had strategic value for the northern armies.

The steam engine was by no means only a weapon of war. In fact, the steam engine made it possible to build vast new factories, which in turn led to urbanization after the civil war. This urbanization was fostered by the huge number of immigrants, whose poverty, illiteracy and heterogeneity made the targets of the most brutal exploitation. The immigrants had arrived by sea on ships driven by steam engines, and continued to cities all over the United States by means of using the locomotives of the day. This led to the development of an industrial working class and the violence and misery of factory life, as depicted by Upton Sinclair and other authors. The steam engine was everywhere in the lives of these newcomers. Pile drivers, steam hammers, and locomotive whistles made endless noise, increased by the noise of freight trains coupling and uncoupling in unending succession.[9]

The use of the steam locomotive in America began around 1830 and continued for a century. It united the country from east to west when the Central Pacific and Union Pacific created the first transcontinental railroad. This achievement had numerous consequences. For one, the bison or buffalo became almost extinct in America, although there had been 60 million bison in North America at the beginning of the 19th century. This decimation was caused by the practice of halting locomotives where bisons appeared, so that crew members and passengers could shoot bisons for the amusement of the railroad's customers.[10]

The locomotive and the railroad business created the first large corporations in the United States and led to the establishment of a few super wealthy families who controlled economics and politics for a century. After the Civil War, the United States changed from an agrarian to an industrial society. Hand labor became obsolete, as mechanical labor by means of the steam engine became the norm. Distances were overcome as never before, as commodities could be produced far from where they were sold. Even time became the province of the steam engine, as the railroads maintained schedules which were imposed on the public, who lived by the time of the arrival or departure of trains. The owners of the railroads moved about the country in luxury trains equipped with the most expensive furnishings. These rail magnates bribed public officials and became more powerful than any elected politician.[11]

The so-called "robber barons" who owned the railroads lent their luxury trains to politicians running for office so that the candidates could harangue the crowds from the back of the train. This practice was common among presidential candidates, like both Roosevelts,Truman, Eisenhower, and Reagan, who used the Magellan luxury train in their campaigns.[12]

Locomotives were also used to manage transportation within cities. The first mechanized form of public transportation in New York City and other urban centers were locomotive pulled cars on elevated tracks. These locomotives were smaller than those used over long distances and were called "dummies," as the machinery was hidden by a full sized car body. These smaller locomotives were also used in factories before the introduction of the motor truck and fork lift. These engines traveled on narrow gage tracks only 18 inches wide.[13]

III.

One of the consequences of the invention of the steam engine was the immigration of thousands of Chinese and other immigrants to the United States in the 19th century.The use of the steam engine as a locomotive is of course dependent on the availability of tracks. In the 1830's, when trains were first introduced to the vast distance between the east coast and the west coast, it became imperative that track had to be laid from coast to coast and across the entire country in all directions. Therefore, thousands of men were employed in the labor of laying track. These were former soldiers, Chinese, Irish and other immigrants, and anyone else willing to undertake this grueling task. The Illinois Central Railroad alone employed ten thousand men to lay 3,800 miles of track between 1852 and 1856.[14]

The work done by these Chinese immigrants, who started in California and moved east, and the Irish, who started in the east and worked west, was indeed grueling. The men who labored at these hard chores had to blast their way through mountains and lay tracks across deserts. They endured terrible winters and overheated summers. They had to defend against hostile Native Americans and deal with political and corporate corruption. These men finally built 381,000 miles of railroads, of which only 167,000 are still used today (2012).[15]

When Chinese men first applied for work on the western railroads, many Americans viewed them as unsuitable for physical labor because their average height was only 5 feet and the weight of these men was only 120 pounds. Yet, it turned out that the Chinese men were willing and able to work from sunup to sunset six days a week for $28 a month. This was later raised to $41 a month when the Chinese risked their lives to blast tunnels through the mountains for the Central Pacific Railroad. To the surprise of their Irish supervisors, the Chinese did not fight, did not strike, and kept themselves clean.

When the Central Pacific Railroad tracks met the Union Pacific's tracks at Promontory Summit on May 10, 1869, and united the country from "sea to shining sea," Chinese immigrants had at least earned half the credit for this singular accomplishment. About 11,000 Chinese had laid the tracks from Sacramento to Utah. Yet, on May 8, 1882, President Chester Arthur signed the Chinese Exclusion Act, which Congress had passed to satisfy the bigotry of so many California citizens.[16]

The Irish immigrants to the United States were mainly responsible for laying the tracks allowing cross country travel from the east coast to the west coast. Nearly 2 million Irish came to the United States during the decade 1830-1840, so that these mostly unskilled workers found employment on the railroads and in the digging of the Erie Canal. These laborers were called trackmen, and like the Chinese of the west coast, worked six days a week from sunup to sundown at laying tracks over the vast continent before them. There were, however, many differences between the Irish and the Chinese. Working in remote and sparsely populated areas, the Irish trackmen lived a dangerous life. The death rate from disease, accidents, and exposure was very high, as was the crime rate. Unlike the Chinese, the Irish fought among each other and against members of other ethnic groups such as the Germans. The fighting between Irish workers consisted mainly of fights based on their origin in Ireland so that those from county Cork, would fight those from county Longford, etc. They also fought law enforcement, which was mainly in the hands of the National Guard and some private railroad police, because police departments hardly existed in this country before the 20th century.

The Irish laborers were not hired by the railroads but by private contractors who garnished some of their wages. These contractors were mostly Irish Protestants who, as in Europe, "lorded it over" the Roman Catholic laborers they hired.[17]

Beginning in 1854, the Irish workers on the railroads went on strike, which was unknown among the Chinese in the western part of the country. Fearing layoffs and unemployment, some Irish laborers sought to gain a monopoly for themselves and those from their home area. This was one of the main reasons for the fighting and the strikes by these immigrants.[18]

IV.

Another consequence of the expansion of railroads throughout the United States was the invention of time zones. Prior to that invention, time was only local. It was noon when the sun was directly overhead, according to "solar time." Every town had an official clock, which was used to set private watches accordingly. Whenever someone traveled to another town he set his watch to suit the time in that location.Once railroads were capable of moving passengers and goods quickly over large distances, the use of solar time became utterly confusing.

Then, in 1878, Sir Sandford Fleming (1827-1915), a Canadian engineer, invented the time zones. Fleming divided the world into 24 time zones, each 15 degrees on longitude apart. Fleming based his time zones on the fact that the earth completes its rotation every 24 hours. Since there are 360 degrees longitude, the earth rotates $1/24^{th}$ of a circle, or 15 degrees, each hour.

This scheme was adopted by American railroads in 1883. Then Congress passed the Standard Time Act in 1918, so that today the U.S., including Alaska and Hawaii, cover nine time zones. Since the railroads were first in adopting these time zones, it is evident that time zones were yet another outcome of the use of the steam engine.[19]

As the railroads became more and more efficient in moving large numbers of people across the country, the population of the western states increased dramatically. Thus, the population of California doubled between 1900 and 1920 and continued to grow more and more thereafter. During those same years, 1900-1920, the population of the United States increased 11% while the western states grew much faster, with California leading with 53%, Washington increasing 37%, Oregon increasing 39%, Nevada 45%, Utah 25%, Arizona 50%, New Mexico 28%, and Texas 20 percent.[20]

Then, in the second half of the 20th century, the use of the steam engine as locomotive began to decline. Replaced by diesel and electric engines, by 1990, steam engines had become extinct in American land transportation as the combustion engine in the form of automobiles and the airplane took its place.[21]

V.

Steam transportation by sea actually preceded steam transportation by land. Nevertheless, steamships were not very useful when first developed because of the need to limit weight and space on a ship. This means that even after it became possible to move a ship by steam, an efficient way of doing so did not succeed until Watt's double acting rotative engine was built. This led to a considerable number of efforts in France, England, and the United States to build a viable steamboat. It was Robert Fulton who finally succeeded in doing so when he introduced regular steamboat navigation in 1807 on the Hudson River.[22]

Although Fulton was able to launch a steamship on a river, it took a good many decades after 1807 before steamships could navigate the oceans. That development depended on John Elder, who was able to build a compound engine in which the expansion of steam takes place successively in a number of cylinders, thereby making a seagoing vessel possible. Yet, Elder based his engine on earlier versions of the compound engine as produced by Hornblower, McNaughton and Wolff.[23]

Therefore, riverboats preceded oceangoing vessels, which did not come into commercial use until the 1830's and beyond. Riverboats were used by the Oregon Steam Ship Co., whose ships were used on the Columbia River and on the Colorado River in the 19th century. The most famous the riverboats were, of course, the steamboats on the Mississippi, the Missouri, and the Ohio rivers. There was a time when more than 1,000 river boats operated on the Mississippi. These steam boats were described by the author Mark Twain in his bookLife on the Mississippi.[24]

The first of these steam boats was the New Orleans, launched in 1811. This boat was a side-wheeler. The Comet was the second Mississippi steamboat, launched in 1813, followed by hundreds of boats ranging from 80 to 140 feet in length. The size of these boats became larger as the nineteenth century progressed, until the Natchez VIII reached a length of over 303 feet.[25]

The boilers of these boats were not safe, so that boiler explosions occurred repeatedly for many years. Some 500 such vessels were wrecked by boiler explosions, killing hundreds of passengers.[26]

Gambling, drinking, and whoring were common on these boats. Most men carried weapons such as guns and Bowie knives. Counterfeiting was also common practice on the Mississippi riverboats, as was the consumption of a great deal of alcohol. While such practices may have been regrettable, it is a fact that many a town at the edge of the great river owed its existence to these river boats. Likewise, the boats existed because they carried people and cargo to all the towns and cities along the Ohio, Missouri, and Mississippi all the way from Minneapolis to New Orleans. The Mississippi had become the fastest and most convenient way to move long distances from one town to another.[27]

Beginning in 1858, some riverboats included theatrical performances or song and dance teams as the entertainment for passengers as well as folks living at the edge of the river. These showboats were developed by Joshuah C. Stoddard, who installed a "calliope" (Greek = beautiful voice) on the Palace and then organized the Steam Boat Music Company. A calliope is a musical instrument powered by steam.[28]

These developments were followed by the invention of the screw propeller, the steam turbine, the electric generator, and systems of power transmission which made it possible to connect the turbine to the propeller shaft. Furthermore, ships were made of iron and steel rather than wood and all of this created the steamer or ocean liner that literally brought millions of "the wretched refuse of your teaming shores" to the United States (this author was one of them). Evidently, the steamship was the product of many minds and many inventors, leading to the conclusion that the "heroic" theory of invention cannot be defended.

The S.S. Savannah was the first steamship to cross the Atlantic Ocean in 1818. This ship still carried sails and did indeed use them during part of the voyage. As steamships improved, they shortened the length of the voyage to Europe from a minimum of five to six weeks to less than two weeks. The steamers also were more predictable as to their arrival time. One consequence of this development was that more temporary migrants came to the United States than before, as a return to Europe was now more feasible.[29]

It took about fifty years for the transition from sail to steam to occur in the Europe to New York immigration movement. This transition took so long because the steamship was not at first economically competitive and because mass migration from Europe to the United States did not develop until the beginning of the 20th century, when 800,000 to one million immigrants arrived on the east coast ports every year. An example of this vast migration occurred in the year 1907, when 1.25 million immigrants came to the United States from eastern and western Europe. The same number came every year until Congress passed the immigration law of 1923, which limited immigration to 3% of the nationality already in the U.S.A. in 1910.[30]

The increase in arrivals was largely due to the persecution of Jews in the Russian empire and to the economic misery of the Italian peasants. Nevertheless, so many of these people could not possibly have come in one year if the size of steamships had not increased dramatically between 1852, when an average of 40 immigrants came on each voyage, until in 1873 about 150 immigrants came on every one of 36 ships making between them fifty-five voyages that year. [31]

Meanwhile, the size of steamships increased as did the number of steamship companies, so that with increased competition the fares on these ships declined. The fare from Glasgow to New York was 8lb. 16s. in 1853, but was only 5lb. in 1875. Similar reductions were the rule from British, German and French ports. [32]

Over the years the size of steam ships increased. In 1831 the SS Royal William's gross tonnage was 540. By 1840 the tonnage of the SS President had reached 2,366. In 1897 the German steamer Kaiser Wilhelm der Grosse had reached 14,349 tons, only to be surpassed by the Bismarck at 56,551 tons in 1914. In 1940, the British steamer Queen Elizabeth, at 83,673 tons, became the largest steamer in the world. Subsequently both British and American cruise lines built ocean liners reaching 225,282 tons when two ships that size were launched in 2009.[33]

The steam engine transformed the navies of the world radically. For the United States Navy, steam meant that the navy, which had been insignificant before 1885, rose to world preeminence after that year. Indeed, the U.S. navy had some successes during the early years of the republic and even during the War of 1812. Thereafter, the American wind driven navy became the object of British ridicule when Puck magazine called it "three mud scows supplemented by a superannuated canal boat."[34]

Beginning in 1890, American warships were the most modern ever built. Instead of depending on sails, the new navy was steam powered and the ships were made of steel, not wood. These new steel ships required new engineering skills, as it was important that the engines and the hull were well integrated. In fact, the engineers who built the new navy were generally recruited from those who had built locomotives, as the ships were armored with specially hardened armored plates. The new naval ships were built by the same producers who were building locomotives along the Delaware River in Pennsylvania and Delaware. These navy yards built 31 battleships between 1890 and 1909.[35]

Between 1910 and 1929, there was a decided increase in the tonnage of battleships being built in American shipyards. Almost all of these were on the east coast. The motivation for this increase lay in part in the "dreadnaught fever" of the European powers, England, Germany, France, and Russia, and in Asia, Japan.[36]

Barbara Tuchman, in her fascinating book The Guns of August, describes in detail the difficulties the battleships of the First World War encountered in the Mediterranean. These difficulties were mostly boiler troubles, understaffed furnaces, and the need to take on coal in different ports, together with the immense exertion by hundreds of sailors in transferring coal from shore or from "coalers" to their battleship. These ships were also limited in the amount of speed they could attain in warfare situations, inasmuch as speed depended on the boilers installed in each ship.[37]

This increase in naval shipbuilding was accompanied by an increase in employment. For example, in 1914, when the United States was not yet involved in the First World War, only 2,178 men worked in naval shipyards. At the end of the First World War in 1919, that number had risen to 11,234, and reached 42,893 in 1943 during the Second World War.[38]

VI.

That increase in employment was not limited to shipyards and/or rail transportation. One of the most important increases in employment resulting from the use of the steam engine was the increase in employment in industry. This use of steam in American factories was visible to anyone ever working or visiting there, as industrial steamers did the work later done by trucks and forklifts. Narrow gage tracks only 18 inches wide were laid through steel mills and brick yards, foundries and coke ovens. These diminutive steamers also serviced stone quarries, gravel pits, sawmills, or any other industry needing to transport almost anything a short distance. The small steamers were also used in the forestry and mining industry. These machines took the place of animal power, so that twenty-one mules and their driver could be replaced by one small steam engine while saving a good deal of money by doing so. These locomotives expelled fumes from burning coal and coke. They also made a great deal of noise, so that work in industrial settings was hard on the laborers, who worked up to 16 hours a day in these factories.[39]

Steam engines precipitated the Industrial Revolution because water is readily available and is inexpensive. Furthermore, steam is clean, is relatively safe, has high heat content, and is easily controlled. As a result, American industry has used steam in numerous ways, including the food and drink industry, where steam is used to shrink wrap meat, depress the caps on food jars, and make corn flakes. Steam is also used to keep chocolate soft so it can be molded, and for heat shrinking a film wrapper on bottles so they become tamper proof. Instant coffee is also produced by steam, and potatoes are peeled with it. Steam is also used in sugar refining.In addition to using it in the food industry, steam is also used in making pharmaceuticals, textiles, and rubber products such as tires. Chemicals are also produced by using steam, and it is also used in oil refining. Numerous other industries have used steam energy to this day.[40]

The use of steam energy did not come easily. As manufacturing with steam increased, labor unrest also increased. An example of this was the development and power of the United Mine Workers of America. Evidently, steam locomotives and steamships depend on the willingness of thousands of miners to undertake the gruesome task of mining coal. This was needed for both locomotives and steamships, as well as industry. Therefore the country depended heavily on the supply of coal to keep just about everything moving. In the nineteenth century, before the development of the combustion engine, coal was the vital substance that allowed the production of steam. This dependence on coal can be seen by looking at the increase in coal production from 1829 to 1899. In 1829, 240,000 tons of coal were mined in the United States. That increased to 15,333 tons in 1859 and to 68,106 tons in 1879. In the last year of that century, 253,741 tons of coal were mined in this country.[41]

A major consequence of the reliance on steam power was therefore the development of the United Mine Workers Union, which was composed of a number of smaller local unions. Beginning in 1885, the new union organized a number of strikes, leading to substantial wage increases. In 1920, John L. Lewis became the president of the UMW, leading to ever greater power by the UMW, as any strike by that union just about paralyzed the country. Lewis was instrumental in organizing the Congress of Industrial Organizations or CIO. During World War II, the government seized the mines because the UMW struck the industry just as coal was badly needed for the war effort.[42]

While coal was used a good deal less as diesel engines and the combustion engine came into use, coal continued to be important to American society inasmuch as one half of all electricity in the United States is produced by means of coal.

VII.

Steam power also gave rise to communism, as defined by Karl Marx (1818-1883), the founder of the communist party and the author of Das Kapital, or Capital, undoubtedly one of the most influential books ever written. Karl Marx lived during the age of the steam engine and was therefore influenced by what he saw in the German, French, and British factories every day. It was principally the British factory system that influenced Marx because he moved to England in 1849 after he was expelled from Germany and France because of his political writings. In England, Marx and family were supported by the German immigrant factory owner Friedrich Engels. This allowed Marx a good deal of insight into the factory system at the time, which included all the features of the industrial revolution as initiated by the steam engine.[43]

The British model of running factories was imitated in the United States, not only because the English language is also used here, but also because a good number of American factory owners were born in England.

In 1832, a member of Parliament, Michael Sadler, investigated the working conditions in British factories. These conditions also existed in the United States and were labeled "a picture of cruelty, misery, disease and deformity etc."Factory owners disputed the report Sadler presented to parliament in 1834. Yet the historians Leigh Hutchins and Amy Harrison regard the Sadler report a valuable collection of evidence regarding factory conditions in the 19th century.[44]

Children were the most abused among the factory workers of the 19th century. Those who described the treatment of children in the factories of England and the United States agreed that children were forced to spend their young years in surroundings of the "utmost immorality and degradation."Factory owners were looking for cheap labor and used children as young a six years for slave labor. Between 1821 and 1850 in some factories in England and America, sixty percent of the factory labor force was children aged ten. Another third of the factory labor force in those years were only eight or nine years old.[45]

Children were also used as chimney sweeps because they were small enough to push themselves up narrow chimneys and scrape the soot off the chimney walls. Observers of factories which employed children described them as places of sexual license, foul language, cruelty, violent accidents, harsh punishments, low wages, and unhealthy conditions. The steam engine made it possible for factories to establish themselves in areas which were not near water. Employers then hired children from poor families to work in these hideous factories.[46]

Not all factories employed children. However, conditions for adults working in these factories were equally horrible, as men and women slaved sixteen hours a day six days a week for so little pay they could hardly feed themselves.

It is well known that Karl Marx explained the exploitation of "the working class" by arguing that the capitalists own the means of production such as land and factories while the exploited worker owns nothing but his labor, which he sells cheap in order to eat. This leads, according to Marx, to the control of the laboring people by the capitalists. Marx further claimed that the value of a commodity arises from the labor incorporated in it and that therefore the laborer is exploited because he earns far less that the worth of the items he has produced.[47]

Evidently, Marx could not have known that the era of exclusive steam power was coming to and end and that much of what he saw would end with it. Marx could not have known that the automobile, the tractor, and the oil industry would change everything. Nevertheless Marx was not entirely wrong in his analysis. Although an enemy of democracy and freedom for the individual, he contributed mightily to our understanding of social stratification and other issues in economics, politics, and sociology. After Marx and Engels translated his work into English, some of his American followers became anarchists. Also heavily influenced by the writings of the French author Pierre Proudhoun (1809-1865) and the Russian writers Mikhail Bakunin (1814-1876) and Peter Kropotkin (1842-1821), anarchists in America and Europe assassinated numerous heads of state in order to promote the classless society and other Marxist aims designed to end the exploitation of the laborers in the factories governed by the steam engine.[48]

Marx was by no means the only 19th century believer in a utopian future. The invention of the steam engine and its diffusion across the United States and Europe led a number of writers to predict that the machine would eventually replace all human power and lead to a perpetual paradise in which no one needed to work and all things would be produced by machines.

Robert Owen (1771-1858), a Welsh socialist and agnostic, believed that a world without labor was at hand, as did John Etzler, the author of *Paradise Within the Reach of All Men*, in which he predicted that hard labor and poverty will disappear as the steam engine and other yet to be invented devices will lead to a paradise on earth. Even Jonathan Edwards, the principal proponent of "The Great Awakening" theology, believed that a paradise on earth was about to begin by means of power derived from machines.[49]

A number of 19th century historians argued that the difference between primitive and advanced civilizations could be measured by the manner in which energy, or the ability to do work, had developed in each culture. It was claimed by these observers that some cultures had reached an "energy stage," which was the prerequisite for the development of an "advanced" civilization, as contrasted with a "simple" civilization, whose energy depended entirely on human and animal power.[50]

Wealth and abundance of all resources and no more human labor were viewed as the certain outcome of the new industrial age as envisaged by Hegel, who had taught that all history is an inevitable march to growth and a final paradise. That paradise was foreseen by John Adolphus Etzler, a German immigrant who came to the United States in 1833 and published a book *Paradise Within the Reach of All Men*, in which he wrote that increased steam power would soon eliminate all drudgery and create "a society in which everything desirable for human life may be had by every man in superior abundance."[51]

These views were indeed exaggerated. Nevertheless, the inventors of that age became heroes because their inventions allowed innumerable men to substitute steam power and later coal, oil, electricity, and atomic energy for muscle power. It was widely believed in the 19th century that inanimate machines would replace all human effort. The German philosopher William Ostwald viewed the history of civilization "as the history of man's advancing control over energy."[52]

The 19th century ideas concerning the eventual glorious outcome of the use of energy were largely produced before physics became an independent science. The inventors of the 19th century were convinced of the efficiency of the steam engine in converting heat to work. Among these leading writers of the day was Sadi Carnot of France, who wrote a whole book on how to maximize the efficiency of steam engines. In addition the Americans Joseph Henry, James

Prescott Joule, and the Englishman Michael Faraday advanced science and industry and provided additional arguments for the utopian visions so common in their age.[53]

In addition, journalists exaggerated the impact of various inventions and discoveries, and promised the public all kinds of utopian results which were not realistic but sold newspapers.

As the steam engine and other devices became the means of production and factories arose in which these machines were used, it became more and more evident that these power producers did not necessarily reduce human labor. Instead, factories using steam, and later electricity, became sweatshops in which the poor were exploited in a most brutal fashion. The factories did indeed create more wealth for their owners. But the power driven machines replaced jobs held by those who then had to work in factories as much as sixteen hours, six days a week to earn a pittance. The factories also became a means of control by which owners forced workers to speed up production as they enriched themselves. Moreover, the physical conditions in factories damaged the health of those who worked there, as coal smoke diffused in the atmosphere the laborers breathed and destroyed the lungs of men, women, and children.[54]

Even as it was obvious that the steam engine led to the growing power of the rich, there were those who believed that the steam engine would eventually eliminate distinctions of rank and money because the steam engine would create cheaper goods available to everyone. This has become partially true in the 20th and 21st centuries, although social stratification remains as prevalent as ever. Surely no one can deny that control of physical power determines the distribution of wealth in the United States. Nevertheless, it is also true that the manufacture of automobiles, washing machines, television sets, and a host of other devices has become cheap enough so that these inventions are accessible to almost all Americans, including the relatively poor.

VIII.

Two sociological processes operate to secure the acceptance of new inventions in any culture. One of these is diffusion. Diffusion means that the new idea is spread from place to place until it may become worldwide. That is what happened to the steam engine, which not only transformed America but also had far reaching effects in European and other countries. The second process is called transmission, and relates to the delivery of a new invention from one generation to the next.. The steam engine was of course sold overseas by Watt and others. In addition it was copied by so-called "pirates." These manufacturers distributed the steam engine all over the European continent, including Russia, France, Germany, the Netherlands, and Spain. At first, the importation of the steam engine had little effect on the people in these countries. It took almost fifty years for the steam engine to "catch on" in countries not accustomed to technological changes. Generally it takes a number of years for a new invention to be acceptable to those accustomed for years, even centuries, to work in a traditional fashion. It is the culture base, that is, the state of technological knowledge available in any one culture, which makes the acceptance of a new invention such as the steam engine possible.[55]

Any new machine or device imported from another culture cannot find acceptance in a different culture unless it is understood by those expected to use it. At a minimum, this means that those seeking to benefit from a formerly unknown machine must have available skilled mechanics capable of maintaining and repairing the newest invention. Where there are no people who can deal with the inevitable breakdowns of a newly imported device, the use of the new machine will not succeed. This is also true of people who have no knowledge of science and industry and who view with suspicion anything they do not know. The culture base can expand only if in the first place the very existence of a new device comes to the attention of a foreign culture. That was certainly true of the steam engine, which was finally introduced to Europe after personal contacts between Americans and Europeans made it possible to persuade Europeans to invest in this new development.[56]

Transmission of the steam engine to the generation succeeding Watt is best illustrated by the development of the Corliss steam engine in 1870. George Corliss (1817-1888) improved the stationary steam engine to such an extent that it became the means by which an entire American industry developed manufacturing enterprises of numerous kinds throughout the United States. Iron, steel, and rails were produced with the aid of the Corliss steam engine, leading to population growth in the United States and in Europe.The evidence for this population growth was that the populations of American counties in which many Corliss engines were employed grew a good deal faster than those in which water mills were used. In fact, Rosenberg and Trajtenberg have shown that communities with establishments using the Corliss engines grew twice as fast in the 19th century than those using other steam engines. The stationary Corliss engines were expensive and were evidence that a good deal of capital had been invested in them. Then, those "counties with more capital intensive industries attracted more population."[57]

Prior to the development of the steam engine, manufacturers who depended on water had to locate where waterpower was available. The Corliss steam engine and others freed manufacturers to locate anywhere. Once located, manufacturers using Corliss engines, which were most powerful, attracted other manufacturers, and thereby increased population in these factory towns. By contrast, communities with a number of water mills declined in population during the nineteenth century. More Corliss engines meant more textile and metallurgy establishments, thereby encouraging more population growth.[58]

Population growth was of course also associated with the steam engine, as used by oceangoing vessels, which brought millions of immigrants to this country, and by locomotives, which distributed hundreds of thousands across the states all the way to California.

IX.

The decline of the steam engine came gradually, with the advent of the Diesel engine. That engine was invented by Rudolf Diesel (1858-1913), a German engineer born in Paris, France, to German parents. He alternately lived in England, Germany, and France, and drowned in the Atlantic Ocean when he fell or was pushed overboard in 1913.

The diesel engine allowed ships and trains to operate more efficiently with oil, not coal. It was perfected in 1897 and was first produced in the United States when Adolphus Busch, the cofounder of the Anheuser-Busch brewery, invested $1 million in the Diesel enterprise.[59]

The invention of the diesel engine spelled the end of steam locomotives sixty years later. Yet, those who built these ever larger engines did not see the decline coming. In fact, during the First and Second World Wars, steam locomotives became ever larger and were used with great success in the war effort. During the Second World War, passenger trains were almost entirely used to transport troops and civilian use of trains was severely restricted.

After the war, in 1945 trains were once more at the service of civilian passengers. These new luxury trains became famous as the "rich and famous" traveled cross country on the Twentieth Century Limited, one of the "great steel fleet" operated by the New York Central Railroad. Not only the east coast railroads, but also Midwestern and Western railroads, invested heavily in new trains, including observation domes, fancy dining cars, and Pullman sleepers. Most extreme in this effort to attract more and more customers was the Chesapeake and Ohio Railroad. Yet, by 1947, a number of railroads were having difficulty filling the seats on their trains, as air travel was beginning. Moreover, by 1948 a few trains had been Dieselized and streamlined. The Hiawatha, a streamlined diesel powered train, traveled from Chicago to the west coast in 1948, as General Motors displayed its new "Train of Tomorrow," with a diesel engine along with diesel powered freight trains. Meanwhile steam engines continued to be produced and sold to American railroads. These steam engines were greatly improved over what was available before the Second World War, as tradition demanded that steam be used despite the obvious diesel advantages. A few railroads had the resources to build their own locomotives even as diesel became the standard in the industry.[60]

The end of the steam locomotive came in 1960. Through the decade of the 1950's this was hardly visible, as the United States was then still a railroad country. During those years, both steam and diesel transported a large number of passengers and freight across thousands of miles, even as air travel and the combustion engine were gaining more and more acceptance. Yet, even in 1950, airline passenger miles exceeded rail travel for the first time. Ten years later, regular passenger steam travel came to an end. Indeed, the steam engine still has other uses. But the era of the great trains is no more, as oil has become the fluid by which America lives and on which 310 million people depend.

The decline of the steam locomotive and rail transport generally is best understood by looking at the railroad mileage between 1930 and 1950 when 250,000 miles of railroad tracks were available in the United States. By 1950 the system was downsized to 224,000 miles and by 2000 only 144,500 miles remained. In 2008 there were still 140,000 miles of rail tracks in the country.[61]

X.

Until sociologists recognized that inventions depend far more on the culture base than on any one individual, it was common knowledge that a few geniuses were responsible for technological and scientific "progress." Yet, the evidence suggests that few advances are the exclusively the work of one man. Instead it is more reasonable to hold that Watt, Fulton, and Edison were born at a time when their contributions promoted the inventions attributed to them even when many others were also involved. The best proof that the culture base, and not any one person, leads inevitably to the next invention, is that in numerous instances more than one person invented the same thing.

Inventions are conditioned by social developments, not by one individual. Few advances are the exclusively the work of any one man. Instead, technological advances are the product of slow accretion. It is popular to believe that Bell invented the telephone and Fulton the steamship and Watt the steam engine. Indeed, each of these men contributed mightily to the eventual success of the technology to which they contributed.

The steam engine is only one example of the conditions leading to new technology. The steam engine came into being when English coal miners needed power to pump water out of coal mines. Prior to the end of the 17th century, this was not needed, as firewood was plenty and the population small. Deforestation made it necessary to use coal.[62]

Likewise, other inventions became the products of the culture base at the time their inventors recognized the means of solving the problems which the needs of their day presented to them. This was true of the steam engine and later of the combustion engine, which was already in the making as steam gradually declined.

Notes

1. Rodney J. Law, *James Watt and the Separate Condenser,* (London: Science Museum Monograph 46, 1969).

2. Augustin L. Taveau, "Modern Farming in America," (Washington DC: *Report for the Committee on Agriculture for 1874),* pp. 282-283.

3. John Fowler, "On Cultivating by Steam," *Journal of the Society of Arts,* vol. 4, (February 1856), p. 167.

4. Reynold M. Wik, *Steam Power on the American Farm,* (Philadelphia: The University of Pennsylvania Press, 1953), pp. 93-96.

5. U.S. Department of Labor, Bureau of Labor Statistics.

6. Ralph Kelly, *Matthias W. Baldwin: Locomotive Pioneer,* (New York, The Newcomen Society, 1946), p.15 and p. 27.

7. John H. White, "Septimus Norris and the Origin of the Ten Wheel Locomotive," *Technology and Culture,* vol. 9, no.1, (January 1968), p. 55-62.

8. Reginald C. McGrane, *The Panic of 1837,* (New York: Russell and Russell, 1965), pp. 40-42 and 92-93

9. Upton Sinclair, *The Jungle,* (New York: Modern Library, 2002).

10. Phillip Guedalla, *The Hundred Years,* (New York: Doubleday, Doran & Co. 1937), p. 42.

11. Alan Trachtenberg, *The Incorporation of America,* (New York: Hall and Wang, 1982), pp. 57-59.

12. Stewart Holbroook, *The Story of American Railroads,* (New York: Crown, 1947), p.433.

13. John H. White, "Industrial Locomotives: The Forgotten Servant," *Technology and Culture,* vol. 21., no. 2, (April 1980), p. 209.

14. Frederick L. Paxson, "The Railroads of the Old Northwest," *Transactions of the Wisconsin Academy of Sciences,* Vol. 17, no. 2, (1911), pp. 269-273.

15. Sean McCollum, "The Muscle that Built the Railroad," *New York Times,* (September 1, 2003).

16. David Howard Bain, *Empire Express: Building the First Continental Railroad,* (New York: Viking Press, 2000), pp. 207-494.

17. George W. Potter, *To the Golden Door: The Story of the Irish in Ireland and America,* (Boston: Little Brown, & Co., 1960), p. 340.

18. Herbert Gutman,"The Worker's Search for Power," In: H. Wayne Morgan, Editor, *The Gilded Age, 2nd Ed.* (Syracuse, The University of Syracuse Press, 1970), pp. 31-53.

19. Lorne E. Green, *Chief Engineer: Life of a Nation Builder,* (Toronto: The Dundurn Press, 1993).

20. Daniel J. McGanney, Frederick B. Whitman and David A. Hill, "Western Railroads," *The Analyst's Journal,* vol. 8, no.4, Proceedings: Fifth National Convention, (August 1952), p. 28.

21. Philip Atkins, *Dropping the Fire: The Decline and Fall of the Steam Locomotive,* (Bedfordshire, U.K.: Irwell Press, 1999), p.102.

22. S. C. Gillian, *Inventing the Ship,* (Chicago: Follett Publishing, 1935), pp. 71-73.

23. Edgar C. Smith, *A Short History of Naval and Marine Engineering,* (Cambridge: University Press, 1938), p.175.

24. Mark Twain, *Life on the Mississippi,* (New York: Signet Classics, 2009, original 1883)

25. Louis C. Hunter, *Steam Boats on the Western Rivers,* (New York: Dover Publications, 1993).

26. James T. Lloyd, *Lloyd's Steam Boat Directory and disasters on the western waters,* (Philadlephia: Jasper Harding, Publishers, 1856), p.41

27. George H. Devol, *Forty Years a Gambler on the Mississippi,* (New York: Henry Holt, 1926), pp. 22-40.

28. Philip Graham, *Showboats: The History of an American Institution,* (Austin: Universityof Texas Press, 1951), p. 78.

29. Gunter Moltmann, "Steamship Transport of Emigrants from Europe to the United States 1850-1914" IN: Klaus Friedland, ed., *Maritime Aspects of Migration,* (Cologne: Bohlau Verlag, 1989), pp.309-320.

30. Ralph Tomlinson, *Population Dynamics*, (New York: Random House, 1965), p.41.

31. Moltmann, "Steamship Transport etc.," p. 311.

32. Terry Coleman, *Going to America*, (New York: Pantheon Books, 1972), p.238.

33. http://en.Wikipedia.org/wiki/Passenger_ship

34. George T. Davis, *A Navy Second to None: The Development of Modern American Naval Policy*, (Westport, CT: Greenwood Press, 1971), p.19.

35. Julius Grodinsky, *Transcontinental Railway Strategy, 1869-1893*, (Philadelphia: University of Pennsylvania Press, 1962).

36. William D. Walters, "The Geography of the European Dreadnaught Race," *Geographic Bulletin*, vol. 34, no.1, (1992), pp. 45-58.

37. Barbara Tuchman, *The Guns of August*, (New York: Dell Publishing Co. 1963), pp. 176-177.

38. No author, "Norfolk Naval Shipyards Portsmouth, Virginia," http://www.globalsecurity.org/military/facility/norfolk_sy.htm

39. John H. White, "Grice and Long: Steam Car Builder," in: Jack Salzman, ed. *Prospects: An Annual of American Cultural Studies*, (New York: Burt Franklin & Co., 1976), p. 36.

40. No Author, "Team Partnership: Improving Steam System Efficeincy," (U.S. Department of Energy, 2000), pp. 8-9.

41. U.S. Department of Commerce, Bureau of the Census, *Fourteenth Census of the United States*, vol. XI, *Mines and Quarries*, 1913), pp. 258 and 260.

42. Morton S. Baratz, *The Union and the Coal Industry*, (Westport, CT: Greenwood Press, 1983), pp. 45, 78, 80.

43. Jiri Marek, "Marxism as Product of the Age of the Steam Engine," *Studies in Soviet Thought*, vol. 32, no.2, (August 1986), pp. 155-161.

44. B. Leigh Hutchins and Amy Harrison, *A History of Factory Legislation*, (London: Routledge, 1903), p. 34.

45. David Keyes, "Industrial Revolution was Powered by Child Slaves," *The Independent*, (August 2, 2010), p. 18.

46. Edward P. Thompson, *The Making of the English Working Class*, (New York: Vintage Books, 1966), p. 307.

47. Abram L. Harris, "Marx on Capital and Labor," in: H.C. Harland, ed., *Readings in Economics and Politics*, (New York: Oxford University Press, 1961), pp. 144-145.

48. Jiri Marek, *Marxism*, p.162.

49. C. C. Goen, "Jonathan Edwards: A New Departure in Eschatology," *Church History*, vol. 28, (1959), p. 38.

50. William R. Burch, "Resources and Social Structure: Some Conditions of Stability and Change," *The Annals*, vol. 389, (1973), p. 27-34.

51. James Ballowe, "Thoreau and Etzler: Alternative Views of Economic Reform," *Midcontinent American Studies Journal*, vol. 11, (1970), pp. 20-29.

52. William Otswald, "The Modern Theory of Energetics," *The Monist*, vol. 17, (1907), pp. 501-511.

53. John Desmond Bernal, *Science and Industry in the Nineteenth Century*, (London: Routledge, 1953).

54. Edward Pessen, *Most Uncommon Jacksonians*, (Albany, NY, State University of New York Press, 1967), part III.

55. L. T. C. Rolt, *Thomas Newcomen: The Pre-history of the Steam Engine*, (London: Newton Abbott, 1963), pp. 58-73.

56. Eric Robinson, "The International Exchange of Men and Machines," *Business History,* vol 1, (1958).

57. Nathan Rosenberg and Manuel Trajtenberg, "General Purpose Technology at Work: The Corliss Steam Engine in the Late 19th Century, *"The Journal of Economic History,* vol. 64, no.1, (March 2004), pp. 83-95.

58. Ibid., pp. 87.

59. Morton Grosser, *Diesel, the Man & the Engine,* (New York: Athenaeum, 1978).

60. Jeff Schramm, "Dropping the Fire: The Decline and Fall of the Steam Locomotive," *Technology and Culture,* vol. 42, no. 2, (April 2001), pp. 368-369.

61. Brian Weatherford, *The State of U.S. Railroads,* (RAND Corp., 2008).

62. Paul Hanley Furfey, "Steam Power: A Study in the Sociology of Invention," *The American Catholic Sociological Review,* vol. 5, no.3, (October 1944), pp. 143-153.

Chapter Two

The Automobile (The Internal Combustion Engine)

I.

It has been widely recognized that the development of the automobile at the end of the 19th century and the beginning of the 20th century changed the United States radically. This is true because the car has many purposes and consequences. Among these were the economic effects which not only led to the employment of thousands of Americans directly involved in building cars, but also led to the expansion of the steel, rubber, and oil industries. In addition, a national system of highways was created, leading to the birth of a major industry after the federal government funded highway construction beginning in the 1920's. These highways then led to the development of suburbs, motels, roadway restaurants, gas stations, garages, and numerous other businesses arising from automobile travel, such as bus lines and advertisements alongside major highways.[1]

This is best understood if we compare the miles of paved roads in this country between a number of census years beginning in 1900. In that year, there were only 144 miles of paved roads. According to the Federal Highway Administration, that number had risen to 2,734,102 in 2008.In 1900, 4,192 cars but no trucks or buses were built in the United States. In 2010 there were 254 million automobiles of every kind in the hands of 196 million American drivers.[2]

The use of the automobile was largely responsible for the breakdown of the distinction between rural and urban life. This means that the users of automobile technology became the agents of social change. Thus, urbanites drove into the countryside and saw first hand how "the other half" lived. Rural Americans drove to the cities to go shopping and be entertained. That entertainment also included sexual escapades in cars, which became popular. For example, a 2003 survey indicates that 40% of men and 16% of women aged 40 to 49 say that they have had sex in cars. In England, one in four respondents reported having had sex in a car.[3] This indicates that the car made it possible to escape he puritan "morality" prevalent in the United States until the 1920's, particularly since children could now escape parental supervision and make the car a "bedroom on wheels."

II.

American history teaches us that the development of the automobile was first felt by rural America because during the years of the initial development of auto travel many Americans still lived on farms or in small towns and villages.In the 1890's, more than 29 million of 63 million Americans lived on farms, constituting 46% of the American population.[4] In addition, another 18% lived in small towns of less than 2,500 inhabitants. This was still true in 1900, when 60% of Americans were rural inhabitants, and even cities were small in comparison to American conditions in 2010. Toledo, Ohio, was larger than Los Angeles in 1900, and the combined population of America's ten largest cities was only nine million.[5]

All this changed rapidly beginning with the 20th century, when farm families first bought automobiles in ever increasing numbers. Consideration was given to the evident advantages of the auto over the horse. Thus, a seven mile trip from a farm to the nearest town by horse and wagon would take most of the day, and was therefore undertaken as infrequently as possible. Even the early automobiles could travel far faster, and make it possible to get repairs for farm equipment in a short time or to drive a sick person to a hospital in far less time than had ever been possible before.[6]

These advantages were not at first evident. Instead, when the so-called "motor car" first appeared in rural areas of the country, driven by wealthy city folks, it aroused a good deal of hostility among farmers. Cars were labeled "devil wagons" because of their "dangerous speed." The new automobiles also had an effect on farm animals. Horses often reared up as the cars made a good deal of noise as they approached, while chickens, which were allowed to cross the roads freely, were often killed by the inexperienced drivers.[7]

In the early years of automobile use, many farm women complained that the reckless auto drivers prevented them from using their horse-drawn buggies, as the horses were afraid of the noise and the very sight of the new invention. The U.S. Department of Agriculture reported that it had become dangerous for women to drive horse and buggy on country roads.[8]

Rural people, not accustomed to the sight of automobiles and their drivers, were often frightened of both the cars and their owners. The cars made a great deal of noise and the drivers would wear goggles and "dusters." The sight of the people wearing goggles appeared un-human or foreign to many a farmhand, who ran from people with such attire, screaming in fear.[9]

Early automobiles were unreliable. They broke down frequently and needed to be pulled out of mud or snow by horses provided by farmers, who felt confirmed in their rejection of the "devil wagons."Furthermore, the drivers were generally upper class city folks who seemed obnoxious to farmers and small town Americans. In addition, the car, which could travel faster and further than horse drawn wagons, threatened the institutions farmers used all their lives. Automobiles made consolidated schools possible, so that the one room schoolhouse together with its primary face-to-face friendships became obsolete. The car also disrupted churchgoing on Sundays as car owners would visit relatives and friends living within range of the auto. Church attendance declined as the need to see other people was better satisfied by long distance driving than walking to church to see the same villagers again and again.[10]

Several state legislatures were so opposed to the use of the automobile that West Virginia and Pennsylvania passed laws seeking to ban autos altogether. In Vermont it was required that a person carrying a red flag walk ahead of any car, while other states enforced legislation demanding that cars slow down or stop for horse drawn vehicles. Speed traps were set up and legislatures refused money for road improvements. Between 1902 and 1907, some farmers attacked motorists by shooting them or beating them. Others threw stones at motorists or placed obstacles in their paths.[11]

In some rural areas farmers dug ditches across roads or weakened bridges to entrap motorists. Others plowed up roads which they themselves had built, or used rakes, glass, pickaxes, barbed wire, and other devices to make auto travel hazardous. An Anti-Automobile League was formed to oppose the car, and in Indiana, a vigilante committee organized to deal "justice" to irresponsible drivers.[12]

Even in the early years of auto production, not all farmers were opposed to its use. Beginning with 1908, when the Ford Model T had been designed, opposition gradually changed to support, as the new car was high off the ground and had other features making it useful in rural areas without paved roads. Eventually farm publications like The Rural New Yorker and Wallace's' Farmer supported cars that could be used in rural areas, such as touring cars with removable back seats which could be converted into trucks. As roads were improved and wealthy city people visited farm communities and spent money there, tourism thrived in hotels and restaurants, and rural repair shops developed. Finally, the 1920 census reported that 30 percent of farm households owned an automobile, as opposition to it came to an end.[13]

Because the automobile could be used to power a number of devices needed on farms, it became more popular than ever in rural communities. Farmers began to use it to power washing machines, but also a number of farm implements such as a corn-sheller, water pump, hay baler, wood saw, cider press and fodder cutter.[14]

Even in the early days of automobile travel, a number of highly significant changes appeared in American social life. The automobile allowed women to get out of the house and increased their mobility in every respect. In 1912, the Ford Motor Company published a pamphlet, "The Woman and the Ford," which suggested that the car was a solution for any woman who wished to bea man and that the car allowed women an ever greater sphere of action.. By 1934, Ford published another pamphlet, "Why Women Prefer the Ford V8," which told women that the car was beautiful and in good taste.[15]

It also allowed children to get off the farm and everyone to widen their horizons. The automobile also allowed extended families to stay in touch with one another. These features allowed auto manufacturers to market the car in terms of family values.[16]

The author is indebted to Ronald Kline and Trevor Pinch for their discussion of "The Automobile in the Rural United States," (Technology and Culture vol. 37, no. 4, October 1996).

III.

In the course of the 20th century, the automobile developed from an expensive toy for the rich into a necessity in most developed countries. It became the standard vehicle for the transportation of goods and people, and thereby led to immense changes in the lives of millions all over the world.[17]

The positive consequences of this development cannot be overestimated. Nevertheless, there are those who have listed a number of negative results of the automobile age which, taken together, make it appear as if the automobile is a curse upon mankind.

It is most certainly a positive development that the automobile has allowed far easier access to places at one time remote and inaccessible. Even as that access became realized more and more, traffic congestion often negated the advantages of auto travel. Furthermore, the automobile led to the movement of millions into remote suburbs, making commuter travel more and more common and contributing to the kind of gridlock from which so many large city residents suffer. This is particularly the case in Latin America.[18]

Employment has been immensely affected by the automobile. There is a vast network of paved roads in America which must be maintained constantly. This network employs 79,730 construction workers who earn a mean salary of $37,290 a year. This large employment became possible as the number of paved roads increased.[19]

The use of the automobile has of course led to numerous changes in American life, including employment, shopping patterns, social interaction, and city planning, as well as reductions in the use of horses and railroads. These changes became immediately apparent once the automobile had entered American life before it was a worldwide phenomenon. This means that as early as 1927, eighty percent of all cars were driven by Americans, as the auto industry had become the largest in the nation.[20]

In 2010, the total employment in the American auto industry had declined to about 370,000 from a high of nearly 650,000 in 2000. This decline was due to the recession which encompassed all American industry in that year, but also to the ever growing "outsourcing" trend which sent American jobs overseas and to Mexico.[21]

The use of automobiles has also reduced railroad passenger travel.Before the Second World War, numerous railroads offered service to almost all cities and towns in the United States and the largest lines employed such high speed trains as the Hiawatha, The 20th Century Limited, and The Empire Builder.These trains would travel 1,000 miles in only 13 hours nonstop.Yet, after the Second World War, passenger traffic dropped quickly. At first, the railroads made numerous improvements and increased the comfort on their trains. But by 1960 it became evident that nothing could stop the decline in rail passenger volume, so that the railroads finally ceased all passenger service as the U.S. government introduced Amtrak on May 1, 1971. Passenger railroading is not profitable so that Amtraksurvives only because of government subsidies.

All this came about because Americans preferred to drive their own automobiles, although the use of airplanes also contributed mightily to the decline in passenger travel on railroads.[22]

IV.

Although in the second decade of the 21st century some Americans are leaving the suburbs and are returning to the cities which their grandparents left, it is still true that in 2010 41 percent of Americans lived in cities and 59 percent lived in suburban or rural areas. Numerous suburban families have moved out of the older suburbs and found housing in the so-called "exurbs" which are farther from the city limits than the old suburbs but are nevertheless viable because stores, shopping centers, factories, and other employment opportunities have also moved to the exurbs.[23]

As more and more people moved into the suburbs, industry and commerce also followed because the automobile made that possible. The best example of this development was the creation of Levittown, New York by Alfred and William Levitt in 1947. They built a planned community in the Town of Hempstead, in Nassau County on Long Island. The town has a population of about 53,000. In addition, the Levitts built Levittowns in Pennsylvania, New Jersey, and Puerto Rico. The houses were about 60 feet apart and were mass produced using uniform and interchangeable parts, allowing the Levitts to build 200 houses a week.[24]

This mass production of houses led Malvina Reynolds to write a song in 1962 called Little Boxes. Reynolds lived in California, where she saw similar developments.

When Levittown first opened, blacks were excluded either by so called "covenants" or by violent reactions to the presence of blacks in the suburban developments. According to Kenneth T. Jackson, middle class whites move to the suburbs because of the affordable housing they found there but also because of racial prejudice.[25]

As office complexes and manufacturers followed the population into the suburbs, cars became essential means of transportation to work, to shopping malls, and to recreational facilities, because public transportation did not follow the suburbanites to where they lived. This in turn led to major traffic congestion in the suburban areas around large cities, as more and more cars were entered into the traffic. Shopping malls permitted suburbanites to avoid big city department stores in congested downtown areas. Nevertheless, most employment continued to be located in cities, so that commuting by car became common for millions of Americans who live in the suburbs. Therefore, traffic congestion has become the norm in America, whose population drives 251 million cars and 7 million motorcycles every day.[26]

As the older suburbs lose their original populations because many homeowners move to newer homes in the exurbs, farther from city limits and depending entirely on automobiles, the original suburbs begin to have all the problems of the cities. The older suburbs include mostly old people with limited incomes, as their children and grandchildren move further from the cities and leave decaying housing behind. Even the peace and security which brought suburbanites to their homes outside the city limits become problematic, as gated communities arise to protect the residents from crime.[27]

Suburbs are of course both geographic and social environments. Because they are easily accessible by automobile, suburbs provide low density housing to those who want to escape the congestion of large cities. Suburbs are also a state of mind, bringing with it a family centered way of life. This came about early in the 19th century, when families gradually relinquished their role as economic units of production to factories. Families then became child centered sources of companionship, and the suburbs the ideal places to carry out these functions.[28]

Suburbs, unlike cities, provide few public places where one can meet strangers, or at least non-family members. In cities there are taverns, bowling alleys, movie houses, and shopping centers, all easily accessible. In the suburbs all of these places can only be reached by automobile, as the suburbs are decentralized, giving rise to the term "urban sprawl." Therefore suburban families need to deal only with each other in isolation or drive or be driven to innumerable activities involving adults and children. For some families, that is too much of an emotional burden, and creates a kind of "claustrophobic" togetherness.[29]

This means that the promise of the suburbs has not always been fulfilled and is at least one reason for the reversal of the trend to move to the suburbs as experienced at the beginning of the 21st century. Another reason for this reversal is that more and more Asian, Hispanic, black, and other poor people moved to the suburbs during the thirty years ending in 2010. Because everyone can afford at least a used or previously owned automobile, a demographic inversion occurred, which was first visible in the census of 2000. Furthermore, the number of married couples in the United States fell to fewer than half of American households, leading to a surfeit of single people in the suburbs. Married couples with children have become a minority in the United States. Most married couples still lived in the suburbs in 2010.However, single people often left the suburbs to seek the amusements, entertainments, and singles activities available in cities but not in "ticky tacky" suburbs. As a result of this "white flight" into the cities, the white population in American cities grew during the decade 2001-2010.[30]

There can be no doubt that the suburbs are car dependent. This dependency is based in the main on the observation that more than half of Americans live in suburbs. Some cities have attempted to reverse this century old trend by "retrofitting" city housing by deflecting growth into already developed areas but not the suburbs. The idea is to give cities a suburban look by making cities more pedestrian friendly by increasing public transportation and decreasing dependency on the automobile.[31]

During the 1960's, when the civil rights and women's liberation movements constantly occupied the attention of Americans, there were those who also attacked automobile transportation as a "rights" issue on the grounds that the car caused pollution, was a symbol of materialistic capitalism, and posed a threat to the environment.

Racism was also invoked, as more and more interstate and other roads were built, some of which cut through black neighborhoods. These roads did indeed displace many established communities, as critics complained of "White roads through black bedrooms."[32]

V.

It is no exaggeration to say that the automobile has become a status symbol, particularly for adolescent Americans. Adolescence (Latin=toward growth) is an American development also known in Western Europe, but by no means universal. In the vast majority of nations, adolescence, i.e. a period of growth located between childhood and adulthood, does not exist. Instead, the majority of cultures move their members directly from childhood to maturity. In the United States and some Western industrial and post-industrial countries, long periods of education create an interval between physical maturity and economic independence called adolescence.[33]

While the traditional "rite of passage" from childhood to maturity was a religious ceremony usually identified as confirmation, the "rite of passage" from adolescence to adult standing has become the right to drive an automobile. That right is confirmed by the issuing of a driver's license. The age at which driver's licenses are issued differs from state to state, although the majority of states allow driving a car at age sixteen. A majority of states issue learner's permits to those who are only fifteen years old, except for North and South Dakota and Wyoming, where such licenses are given to fourteen-year-olds. Full and unrestricted driver's licenses are issued to those age 17 or 18, except for New Jersey, which requires that a fully licensed driver must be age twenty-one. Driver's licenses not only serve to allow driving a car but are also used as means of identification and proof of age. This is particularly important to adolescents who seek to drink alcohol.[34]The driver's license is believed to increase the adolescent's sense of independence and power.Adolescent boys are anxious to be accepted as "men," a status which depends in part of driving a car. In addition, dating girls generally depends on the use of a car, not only because of the image of masculinity associated with driving but also because public transportation is difficult to access outside of New York City and other super sized communities. As to masculinity, the car owner views himself as one who controls technology, particularly if he knows or claims to know how to fix car problems or alter the car by attaching oversized tires or other conspicuous devices.

Among the adolescent boys who were anxious to drive a car because they wanted to date girls are numerous young men who spend their time and money on refurbishing their cars, re-boring cylinders, and "working" endlessly on the automobile, instead of dating as they had anxiously anticipated until they finally own a used car, which then interests them more than the girls.[35]

Adolescents are also known to drive carelessly with excessive speed and often without authorization to use the family car. Boys are more likely to behave in this fashion than girls because male prestige in this country is more often dependent on the display of automobile prowess than is true of female prestige. The automobile is in any event a prestige symbol in American culture for adults as well as adolescents.

While girls gain a good deal more from the clothes they wear than from cars they drive, boys need to "show off" by speeding or even stealing a car, thereby attaining adult standing among their peers. Such behavior also provokes parents to pay attention to boys who feel they are ignored or not sufficiently appreciated by parents. Accidents which result from careless driving force parents to protect the boys involved.

American boys view the automobile as a symbol of masculinity. Masculinity is an issue in the lives of Americans throughout their lifetimes, beginning as soon as a boy is old enough to discern the difference between male a female. Therefore, is all cultures, the symbol of masculinity is highly prized. In American culture that happens to be the automobile.[36]

There are also additional reasons for the use of cars by adolescents as soon as that is legal. In the upper middle class, children, including adolescents, are generally engaged in numerous activities after school, on weekends and during vacations. Driving children to all of these encounters is exhausting and time consuming for parents who are both working so as to maintain the upper middle class status. Therefore, many adults are happy to have their young adult drive as soon as possible in order to rid themselves of that task.[37]

Getting a license to drive means that adolescents spend more time away from their family, although for many the right to drive results in the expectation that the drivers will share more of the family responsibilities such as driving younger children to piano lessons or helping get the groceries home.[38]

Parents seek to exercise a good deal of control over their adolescent children. This control is in part derived from imposing rules on driving the family car, on dating, and on curfews and other prohibitions. Nevertheless, children have a good deal more power than is commonly attributed to them. That means that there are numerous negotiations between children and parents, leading parents to feel that adolescents are particularly difficult as they insist on more and more independence.[39]

Social class is a major determinant of automobile use by adolescents. Social class is of course decisive in all aspects of family life, with particular emphasis on family resources. Evidently, some families can afford to give each child a car at age 16 or whenever the legal driving age occurs. Others share the one family car, with the result that driving becomes a great privilege for poor children, including those whose families have no automobile.

Those adolescents who drive not only free themselves from the constant supervision of parents, they also increase their field of action. The driving young woman or man can access friends in neighborhoods outside their own, can reach places of public amusements such as theaters, bowling alleys, dance halls, and drug related associations. Self-determination becomes more and more possible for those who drive. All this leads to a good deal of parental anxiety concerning their children's safety and the gradual loosening of the parent-child bond.[40]

Adolescents who don't have a car themselves must ask parents to drive the family car. They are therefore obliged to deal with their parents' work and leisure schedule, because the car ultimately belongs to the parents and not the adolescent. In many such families, there is a constant struggle between adults and their children as to the use of the family automobile.

Social class determines a good deal concerning the use of cars or automobiles. Evidently, parents with considerable resources can buy cars for any child sixteen years old or older. This makes a big difference to high school students, who self-segregate into "insiders" and "outsiders" in almost any American secondary school. Clothes and cars are signals of belonging to the "ins" or "outs" among high school students, who mimic adult behavior relativeto social class as found in every American community. A student who has been given a new sports car is instantaneously elevated "above the crowd," even as those who have no such possessions become the outsiders among the high school circle. This behavior by high school students is related to the concept of "conspicuous consumption," as outlined by Thorstein Veblen in 1902.[41]

Those youngsters whose parents can afford to buy them a car and expensive clothes need not deal with deferred gratification as is required by those with lesser incomes. Consequently, those who have the money become the "big wheels" in high school, with the expectation that they will always occupy that role, even into adulthood.

Because driving a car involves numerous expenses besides "filling up" with gasoline, children are often required to pay for insurance, speeding tickets, accidents, and repairs. The money for all of that has to come from the part time employment of high school and college students. Considering that in 2009 the average (mean) income of Americans then employed was only $40,300, it is evident that the majority who earn such incomes cannot afford all the expenses a car brings with it.[42]

Therefore, most adolescent drivers are dependent on gaining access to the family car on only those occasions when no one else in the family wants to use it. Parents can therefore maintain a good deal of control over those who want to drive. This control is of course much more in evidence among those who partially or entirely paid for the children's car. Parents can therefore take away a car as punishment for bad behavior or they can control adolescent conduct by the promise of a car. These circumstances result in children learning the meaning of class in America, a meaning which is vital in producing self evaluation and in projecting one's adult future. Cars and other expenditures become symbols of class membership, as we are defined by what we consume.[43]

Parents always worry about the possibility of accidents and the liability which can come from such accidents while their children drive. This is particularly worrisome when adolescents drive to school and then drive their friends home after school. Those who have a car are the beneficiaries of peer admiration as long as all goes well. If there is an accident, the driver is blamed and his parents are held responsible. That is one of the reasons many a licensed young driver is not allowed to drive to school or to have others in his car with him.Such rules as parents impose on young drivers are frequently gender influenced, as young boys are often accorded more freedom and fewer restrictions than are imposed on girls.This is influenced by the fact that boys are more likely to engage in high risk driving than is true of girls.It is however also influenced by the construction of adolescence on the part of males as a period of testing one's power, a presumably masculine but not a feminine perspective.[44]

For many families, the use of a car by an adolescent is of great help, as the new driver can now chauffeur himself and his younger brothers and sisters to numerous after school activities, thereby relieving parents of endless driving chores. This means that parents can devote additional time to work and leisure time activities.

Once the automobile had become a common and almost universal possession in American families, the question of liability for injuries or damages caused by adolescent drivers became acute. That meant that in the 1920'sand early '30's, the courts had to decide whether or not adolescent drivers, who drove with parental permission, were responsible for accidents caused by them or whether the parents were responsible.

As early as 1904 and again in 1912 some courts held, according to the "agency doctrine," that the owner of the automobile is responsible for any accident caused by a driver who has been authorized to drive the owner's car."[45]

Yet, other courts came to the opposite conclusion. In numerous cases, all decided during the first and second decades of the twentieth century, courts in various states ruled that the owner of the car could be held responsible only if the driver was carrying out a mandate by the owner at the time of the accident.[46]

Various state courts have made different decisions concerning parental responsibility in auto accidents resulting in personal injury or damages to a vehicle. That responsibility is often abused, as teenagers engage in violent behavior including high speed driving, even when they are well aware of the risk they take when driving at 70 miles an hour in a 50 mile zone. The reason for taking unreasonable risks by teenagers is lack of experience.[47]

VI.

In the days before air conditioning became a standard feature of the American automobile, family travel by car was sheer misery compared to travel in the twenty first century. Those were the days when the family would keep the windows open while driving mostly at 55 miles an hour on highways, not "throughways." Because long drives were uncomfortable and exhausting, hotels and "motels" and other enterprises advertised along the roads, such as Stuckey's, who sold pecan candy from roadside stores, Kentucky Fried Chicken, and Cypress Gardens, a one time Florida attraction. Along the road there were also the inevitable "Burma Shave" signs, which promoted their product with such jingles as "Our fortune is your shaven face. It's our best advertising space. Burma Shave"and "The ladies take one whiff and purr—it's no wonder men prefer, Burma Shave" and " We've made grandpa look so trim; the local draft board's after him. Burma Shave." These signs were in use for thirty-six years, from 1927 to 1963, when they were discontinued. Frank Rowsome collected the verses from all of these signs and published them in 1965.[48]

One of the consequences of family drives on weekends, called "touring," were the development of the motor hotel or motel and "auto camps." Between 1920 and 1929, the number of rooms available to guests increased by 29 percent. This included an increase in cabin camps with furnished cottages, which not only allowed more families to tour together but also served as a means of promoting adultery. In the 1930's and beyond, a good number of auto-camps did not require registration because the couples who rented were frequently involved in adulterous relationships.[49]

Long after the Burma Shave signs ceased to be displayed, automobiles became an obsession with many Americans who spent all their spare time in dealing with antique cars.Among those who have such an interest are the numerous antique car clubs who drive and exhibit such old automobiles at every opportunity. Members of these clubs participate in parading these cars, which they preserve and restore at considerable cost. The word "antique" as used in history books refers to the ancient world. Among "antique" car owners it refers only to cars at least eighty years old and dating to the early production of cars by manufacturers who havelong gone out of business.[50]

In the First World War, which took place between August 1914 and November 1918, the automobile became a military resource.This was true for all the belligerents at that time. Historians recognize that Paris would have fallen to the Germans in the first month of the war if it had not been for the requisitioning of all Paris taxis by the French army at the first battle of the Marne, also called "the miracle of the Marne." The Marne is a river only fifty-five miles from Paris. By September 5, 1914, the German armies had arrived there and it seemed certain that they would continue to Paris in a short time. Then the French army reserves arrived in taxis and private automobiles and succeeded in forcing the Germans to retreat.[51]

Of course the French were not alone in using the automobile for military purposes. All the armies fighting the First World War transported supplies and troops to the front and back and also used them as ambulances in the hospital service. At the beginning of the twentieth century, European governments subsidized the development of transport vehicles for military use, as it became increasingly evident that a general war was at hand. This was done in order to insure that all military vehicles were of the same type, since otherwise it was necessary to employ numerous mechanics acquainted with different makes of trucks and cars. Moreover, the parts needed to repair different makes of cars and trucks entailed immense problems which uniformity avoided. This became too much for European car manufacturers, so that the French, the Belgians, and the British ordered vehicles from American manufacturers.[52]

As the First World War bogged down in trench warfare, ambulances were needed to transport the wounded to the rear and to hospitals. These ambulances were equipped with soup kitchens and with liquid nourishment most needed by the wounded soldiers. Bacteriological laboratories and x-ray machines were also transported by trucks to the front and back, and there were additional uses, such as observation, "wireless telegraphy," and others.[53]

During the Second World War, automobiles were also used in the manner of the First World War. However, some major changes were made by both the German and American forces.[54]

The United States developed the "Ford GPA" (G=government, P=80" wheelbase, A=amphibious) and "the tracked landing vehicle" which had been car-boats in the first instance. There is absolute evidence that the United States Marines succeeded in landing on numerous Pacific Islands in the war against the Japanese by the use of these amphibious devices.[55]

The Ford GPA resembled the "Jeep," except that it could swim and cross rivers and lakes.The Germans altered the "Volkswagen" so that it could easily transport troops from ships to shore and across rivers which had no bridges. This vehicle had four wheel drive and could be driven in water.

VII.

The automobile also has a political purpose. While politicians are flying all over the country or a state to campaign for votes, the car serves local needs, even as it was once most useful in the campaigns conducted by Franklin Roosevelt and Lyndon Johnson and others in between. Lyndon Johnson used the car as a means of reaching just about everyone in his congressional district when he first ran for office, and Franklin Roosevelt depended on his car almost entirely as he was seen in his car at all times since he could not walk as a result of having contracted poliomyelitis. There are numerous pictures in old newspapers and magazines showing Roosevelt waving from his car over the more than twelve years he spent in the office of President.[56]

Robert Caro in his book *The Years of Lyndon Johnson: Means of Ascent* describes how Johnson traveled to the remotest Texas farms to introduce himself to the voters, a feat only possible by car.[57]

A free automobile is also one of the "benefits" of holding elected office. Members of the U.S. House of Representatives lease cars at the expense of the taxpayer, who also has to pay for the gas and upkeep of these cars. Many members of the House choose luxury vehicles, as there is no limit on how much a member can spend. Not all members of the House make use of this benefit. Members of the Senate are not permitted to lease cars with public money.

Failure to do so is largely related to a wish to avoid public criticism bordering on ethical issues related to such an expenditure. Constituents have been vocal about extravagant spending on the part of politicians. An example was the denunciation of Representative Maurice D. Hinchey of Ithaca, New York in the Binghamton Press and Sun Bulletin for leasing a BMW 5301.[58]

VIII.

Almost as soon as the automobile had been invented, some drivers engaged in racing one another. The first car races were conducted in France, but soon caught on in the United States, culminating in stock car racing as sanctioned by the National Association of Sports Car Automobile Racing, commonly called NASCAR. Like other major sports, television has catapulted auto racing into a national sport. Racing began as a "blue collar" spectacle, but has now become the province of wealthier Americans, and is also viewed by women. In fact, is has been estimated that 38% of NASCAR fans are female.[59]

On July 22, 1894 the world's first car race took place in France, on a road from Paris to Rouen, a distance of 79 miles. Twenty-five drivers were included in this first race, which was won by Count Jules Albert de Dion.[60]

Auto racing spread from France to other European countries at the beginning of the 20th century until it reached the United States. The first NASCAR race was held on June 19, 1949, at the Charlotte, North Carolina. speedway. It was after the Second World War that car racing became an American sport. The cars used were racing cars with stock appearing bodies. This led to the popular NASCAR 31 race series beginning in 1972 and enjoying an ever increasing number of followers.[61]

IX.

Although the use of automobiles has brought many benefits to the American people and to millions of others all over the world, the negative consequences of auto use are undeniable. The worst of these is the death toll on American roads. This is reported each year by the U.S. Department of Transportation – National Highway Traffic Safety Administration. According to their statistics, 43,510 people died from motor vehicle accidents in 2005. Since then there has been a decline in the number of traffic fatalities in the United States. In 2006, the number of fatalities was 42,708, in 2007 it declined to 41,259, in 2008 37,423 were killed by motor vehicles, and 33,808 died as a result of a car accident in 2009.[62]

Of the 37,261 people killed in auto accidents in 2002, 17,500 died because one or both drivers of crashed vehicles were drunk. Nevertheless, there has been a steady decline in these deaths, not only because the drinking age in most states has been raised to 21, but also because cars were much safer in 2010 than they were ten years earlier; more cars have air bags than ever before and mandatory seat belt laws protect drivers better than was true before such laws were enacted. Police also use sobriety checkpoints, "zero tolerance" laws in many states have made drunk driving more costly, and ignition interlock devices have prevented more drunk driving than warnings could ever achieve. Finally, the appointment of a "designated driver" has made it possible for many a drunken partygoer to get home safely.[63]

X.

The development of automobile use in the United States has been determined by the social construction of technology. The users of automobiles have shaped the automobile technology itself, as different social groups have attached different meaning to the automobile. The manner in which the automobile became a gendered issue best illustrates this proposition. The auto or car became a symbol of power relationships between men and women because the car was almost at its creation designated as an aspect of masculine culture.

The automobile is also a useful index of the growth of the American population. As the number of Americans has grown, so has the use of the automobile, so that the utility of the automobile is declining as the infrastructure, i.e. roads, bridges, parking spaces, all become overcrowded, so that in California and elsewhere, drivers often wait for hours, unable to move. This is called a "gridlock" and refers to the frequent experience in large cities when all movement comes to a halt. This is an example of the law of diminishing marginal utility, which states that each additional unit of use decreases the level of satisfaction available. Hence, as more and more cars use the California freeway system, the average speed of vehicles using that system was only 33 miles an hour in 2000 and nine years later was reduced to 20 to 25 miles an hour.[64]

The growth of automobile usage may be viewed from the Malthusian point of view to the effect that between 1960 and 1980 the American population grew 26% while the number of cars grew at 100%. This immense growth also affects the pollution of American cities, as more and more Americans breathe the poisonous air subject to so much automobile emission.

Over the years, about 60,000 square miles of land has become concrete in the form of roads and parking lots. This is equivalent to the total area of Georgia. The outcome or the cause of this situation is the automobile dependence of the American people fostered by the auto companies, oil producers, and the steel industry, which lobbied the federal government and local governments to build more and more roads and to abolish public transportation.[65]

Summary

The United States changed radically after the invention and deployment of the automobile. One such change was the breakdown between rural and urban life. Another change came about after the initial hostility to the automobile had abated and it was used as a source of power on the farm.

With the decline in the American railroad came the increase in employment in the auto industry, including the construction of roads and the exploration of oil. This led to the growth of suburbs and exurbs, as well as traffic congestion and urban sprawl.

The car became a status symbol, especially among adolescents, who also used it to escape parental supervision. Further, the car had military and political uses. Its downside was and is death on the highways and congestion produced by the law of diminishing marginal utility.

Notes

1. Peter J. Hugill, "Good roads and the Automobile in the United States 1880-1929," *Geographical Review*, vol.72, no.3, (July 1982), pp. 327-349.

2. U.S. Federal Highway Administration, "Highway Statistics 2008," Table HM 12.

3. Mark Edwards, "The Sunny Old World of Sex on Wheels," *Daily Mail*, (February 15, 1992), p. 48.

4. Debra Spielmaker, *A History of American Agriculture*, (Proceedings of the AAAE Research Conference, vol. 34, Minneapolis, MN, 2007).

5. Ben J. Wattenberg and Richard A. Scammon, *An Unexpected Family Portrait of 194,067,296 Americans*, (Garden City, New York: Doubleday, 1965). See also, Mark Baldassare, "Suburban Communities," *Annual Review of Sociology*, vol. 18, (1992), pp. 475-494.

6. Joel P. Kunze, "A Tractor or an Automobile? A 1920's Farm Family Faces a Decision," *Agricultural and Rural Life*, vol. 5, no. 3, (Winter 1991), p. 37.

7. Michael Berger, *The Devil Wagon in God's Country: The Automobile and Social Change in Rural America 1893-1929*, (Hamden, CT: Archon Books, 1979), p.14.

8. Alfred Clough, "Nervous Strain Due to Automobile Driving, *Horseless Age*, (September 23, 1903), p. 324.

9. No author, "Meet Queer People," *Motor Age*, (July 14, 1904), p. 14.

10. F. G. Morehead, "Automobile vs. Country Church," *Technical World Magazine*, (vol. 18, November 1912), p. 208-300.

11. Berger, "The Devil Wagon etc.," pp. 24-28.

12. Reynold M. Wik, *Henry Ford and Grass Roots America*, (Ann Arbor, The University of Michigan Press , 1972), p. 16.

13. Peter J. Hugill, "Good Roads and the Automobile in the United States, 1880-1929," *Geographical Review*, vol.72, (1982), pp. 827-849.

14. Walter Langsford, "What the Motor Vehicle is Doing for the Farm," *Scientific American*, vol.102, (January 15, 1910), p. 50.

15. Virginia Scharff, *Taking the Wheel: Women and the Coming of the Motor Age,* (New York: Free Press, 1991), p. 117.

16. Deborah Clarke, "Eudora Welty's *Losing Battles:* Cars and Family Values," *Mississippi Quarterly,* vol. 62, no. 13, (April 2009), p. 143.

17. John Urrey, "The 'System' of Automobility," *Theory, Culture and Society,* Vol.21, no.4-5, (2004), pp. 25-39.

18. Alan Gilbert, *The Mega-city in Latin America,* New York: United Nations University Press, (2000).

19. Bureau of Labor Statistics, "Occupational Employment and Wages," Washington, DC: United States Government Printing Office, (May 2009).

20. James Jay Fink, *The Car Culture,* (Cambridge, MA: The MIT Press, 1975), p.38.

21. Michael F. Thompson, "Employment and Growth in the U.S. Automotive Manufacturing Industry," *Indiana Business Review,* vol. 85, no.1, (Spring 2010), p. 4.

22. John F. Stover, *The Life and Decline of the American Railroad,* (New York: Oxford University Press, 1970).

23. Jon C. Teaford, *The Metropolitan Revolution: The Rise of Post-Urban America.* (New York: Columbia University Press, 2006), p.7.

24. Levittown Historical Society, *A Brief History of Levittown, New York,* (New York: Levittown Historical Society, No date), p. 3.

25. Kenneth T. Jackson, *Crabgrass Frontier: The Suburbanization of the United States,* (New York: Oxford University Press, 1985), pp. 301 and 327.

26. U.S. Department of Transportation Statistical Records Office, 2010.

27. Anthony M. Orum, *The Centrality of Place: The Urban Imagination,* (Ann Arbor: The University of Michigan, 2004), pp. 18, 51, 69, 73.

28. William Sharpe and Leonard Wallock, "Bold New City or Built-up Burb? Redefining Contemporary Suburbia," *American Quarterly,* Vol.46, No.1, (March 1994):1-30.

29. Philippe Aries, "The Family and the City," *Daedalus,* v. 106, (1977), p. 227-235.

30. Sam Roberts, "Population Study Finds Change in the Suburbs," *The New York Times,* (May 9, 2010), p. A23.

31. Clark Zinni, "Retrofitting Suburbia," *Alternatives Journal,* vol. 26, no. 3 (2000), p. 19.

32. No author, "The Automobile Age," *The Wilson Quarterly,* vol. 10, no. 5, (Winter 1986), p. 76.

33. Ursula A. Falk and Gerhard Falk, *Youth Culture and the Generation Gap,* (New York: Algora Publishing Co., 2005), p.7.

34. 49 *CFR* part 383.

35. Hyman Weiland, "The Adolescent and the Automobile," *Chicago Review,* vol. 9, no. 3, (Fall, 1955), pp. 61-64.

36. Ibid., p. 64.

37. Arlie Hochschild, *The Time Bind: When Work Becomes Home and Home Becomes Work,* (New York: Metropolitan Books, 1997), pp.10-11.

38. Demie Kurz, "Caring for Teenage Children," *Journal of Family Issues,* vol. 23, no. 6, (2002), pp. 748-767.

39. Viktor Gegas and Monica Seff, *"Families and Adolescents: A Review"* In: A. Booth, Editor. *Contemporary Families: Looking Forward, Looking Back.* (Minneapolis, National Council on Family Relations, 1991), pp. 208-223.

40. Laurie Olsen. *Made in America: Immigrant Students in Our Public Schools,* (New York: New Press, 1997), p. 187.

41. Thorstein Veblen, *The Theory of the Leisure Class: An Economic Study of Institutions,* (New York: Macmillan, 1902), pp. 68-101.

42. U.S. Department of Labor, Bureau of Labor Statistics, "Real Earnings September 2010" Table A-1.

43. Mica Nava, "Consumerism Reconsidered: Busing and Power," *Cultural Studies,* vol. 51, no.57, 1991):172.

44. Lyn Mikel Brown, *Raising Their Voices: The Politics of Girls' Anger,* (Cambridge, MA: Harvard University Press, 1999), p. 127.

45. Christain v. Johns, 125 Ga. 977 1904 and Smith v. Jordan, 211 Mass. 269 (1912).

46. Watkins v. Clark, 103 Kans. 629 (1918).

47. Jane Brody, "Teenage Risks and How to Avoid Them," *The New York Times,* (December 18, 2007).

48. Frank Rowsome, Jr., *The Verse by the Side of the Road: the Story of the Burma Shave Signs,* (Brattleboro, VT: S. Greene Press, 1965).

49. Norman S. Hayner, "The Tourist Family," *Social Forces,* vol. 11, no.1, (October 1932), p. 64.

50. Nikasha Dicks, "Families Join GasCar to Share Love of Automobiles," *The Augusta Chronicle,* (October 8, 2009).

51. Barbara Tuchman, *The Guns of August,* (New York: Dell Publishing Co., 1963), p. 487.

52. No author, "The Automobile in War," *Lotus Magazine,* vol.7, no.2, (November 1915), p. 89.

53. Ibid., p. 51.

54. No author, "Amphibious Landing Vehicles of the Second World War," *Car News,* (June 11, 2007).

55. Joseph H. Alexander, "Tracked Landing Vehicles," *World War II,*" vol. 13, no. 4 (November 1998), pp. 1-5.

56. Franklin D. Roosevelt Library, Hyde Park, NY, "FDR Waving from a car."

57. Robert A. Caro, *The Years of Lyndon Johnson: Means of Ascent,* (New York: Vintage Publishers, 1991), Chapter 2.

58. Raymond Hernandez, "What Would you Drive if the Taxpayer Paid?" *The New York Times,* (May 1, 2008), p. 1.

59. Rick Matsumoto, "Stock Car Racing Well on Way to Outgrowing Redneck Roots," *The Toronto Star,* (April 5, 1997), p. E9.

60. Remi Paolozzi, "The Cradle of Motor Sport," *Autosport,* (May 28, 2003), p. 8w.

61. Bill Fleishman, *The Unauthorized Nascar Fan Guide,* (New York: Visible Ink Press), p.6.

62. No author, "Fatalities and Fatal Rates by Quarter," *Traffic Safety Facts,* (Washington, DC, The United States Printing Office, (December 2010), p.1.

63. National Highway Traffic Safety Administration, "Choose Responsibility," Washington, D.C. (2010).

64. Matthew Barth and Kanok Boriboonsomsin, "Traffic Congestion and Greenhouse Gases," *Access,* vol. 35, (Fall 2009), p. 8.

65. Jane Jacobs, *Death and Life of Great American Cities,* (New York, Random House 1961), pp. 341-345.

Chapter Three

The McCormick Reaper and the Agricultural Revolution

I.

As long as mankind has been on this planet, the vast majority of humanity has spent most of its time in the pursuit of food. For thousands of years men roamed the earth as nomads, seeking to feed themselves by primitive hunting and fishing methods, until agriculture made permanent, stable communities possible. That effort to scratch food from the earth by human and animal power lasted for centuries until, in the 18th century, mechanical inventions relieved mankind of that perpetual labor. "In the sweat of your face shall you eat bread" (Genesis 3:19) was an accurate description of most of mankind's fate until Obed Hussey (1792-1860) and Cyrus McCormick (1809-1884) revolutionized agriculture and put an end to human enslavement to the ever recurring need to eat.

Colonial America was of course an agrarian (agr = field in Latin) civilization. That civilization was brought to America by the European immigrants of the 17th century, but also by the Native Americans who had already been in the continent for 20,000 years when the Spanish and the English first arrived. The European settlers raised food, including grain, vegetables, fruit, and meat. They made tools, furniture, and clothes at home, so that all the labor needed to do all that engaged the entire family, including children, day and night, all year, every year of their lives. There was no time to do anything else, so that leisure or schooling or recreation were hardly known to these pioneers.[1]

As the immigrant population moved west from New England into unexplored territory, they brought their agricultural methods with them. These methods were primitive and age old in that farmers relied on their own muscle and the work of their animals. Therefore food production was slow and difficult. Before the invention of the tractor, it took fifty-eight hours to plant and harvest a single acre. After the tractor came into use, it took only three hours to do the same job. Prior to the invention of farm machinery, wheat had to be cut with a mule-drawn header. The wheat was then stacked in piles to be thrashed later. Once the tractor was available, that device did the work of ten horses. Then came the "combine," which cut and thrashed the wheat in one motion with just a fraction of the labor.[2]

These devices gradually became available during the first half of the 19th century. Prior to that, harvesting grain was as old as mankind who, day after day, for three thousand years or more, swung scythes and "cradles" through the grain fields of the world. A "cradle" consisted of a broad scythe with a light frame with four "fingers" attached. This allowed the farmer to throw the grain into a swath so that others could rake it and bind it into sheaves. This and other tasks were endless and hard. Farmers worked in the heat of the sun at labor that took strong backs and powerful arms. It was therefore no surprise that the industrial revolution, begun in England but diffused to America, led to the liberation of millions from farm slavery. Then Cyrus McCormick invented the mechanical reaper in 1831, an invention based on the efforts of his father and others.No invention has ever been the work of only one inventor. All are based on efforts of those who preceded the eventual inventor. Cyrus McCormick benefited most from his father, Robert McCormick (1780-1846), whose attempts to create a reaper seemed to have failed when Cyrus, then twenty-two years old, succeeded. Cyrus McCormick was aided in his work by a slave, Jo Anderson, who was given no credit for his share in creating this important machine.[3]

McCormick demonstrated his invention in July 1831 in a field several miles from his family farm. He improved his reaper for several years and took out a patent on it in 1834. The reaper became a liberator as it released thousands of men from harvesting, as it did the work of five men at the outset. By 1900, the reaper harvested one bushel of wheat in ten minutes, a job which a man with a scythe achieved in three hours.[4]

Before the invention of the reaper, farmers harvested grain by hand. The work was "backbreaking" and slow. Fifteen men could cut, gather and bind ten acres of wheat a day. After innumerable improvements over years, a harvester-thresher or combine will now do the same job in one hour. In fact, the latest model combines will harvest 150-200 acres in one day, depending on weather conditions.[5]

II.

The amount of knowledge concerning any universe of discourse is always evident to more than one person because the culture base inevitably leads to the next conclusion, no matter any particular individual. Indeed, the history of science reveals that every invention and every discovery has more than one father, so that it is sometimes difficult to ascertain who originated a new concept or development. This is true of the invention of the harvester just as it is true of the invention of calculus by Isaac Newton (1643-1727) and Gottfried Leibniz (1646-1716), the invention of the telephone by Antonio Meucci (1808-1889) and Alexander G. Bell (1847-1922), or the discovery of shock waves by Nikola Tesla (1856-1943) and Ernst Mach (1808-1889).

It is therefore not surprising that McCormick was challenged by Obed Hussey (1792-1860), who patented a reaper in 1833. Hussey had already invented a corn husking machine when someone suggested he invent a reaper. He succeeded in 1832, leading him to open a factory in Baltimore in 1838. Subsequently he sold numerous reapers until Cyrus McCormick acquired the patents of others and incorporated these into improvements in his reaper. This drove Hussey out of business and he sold his factory to McCormick in 1858.[6]

In 1855, a reaper invented by John H. Manny of Illinois was exhibited at the Paris Exposition. The Manny reaper was awarded a prize, leading to a lawsuit by McCormick in which he claimed that Manny had infringed on the McCormick patent. Abraham Lincoln was one of the lawyers hired to defend Manny, particularly because the case was to be heard in Springfield, where Lincoln had his law office, and because Lincoln was cheap. When the trial was moved to Cincinnati, Ohio, Lincoln tried to be part of Manny's legal team, but was rejected by the other lawyer, Edwin M. Stanton, who called Lincoln "a damned, gawky, long armed ape and baboon." Lincoln attended the trial, which was won by Manny, although Lincoln was not allowed to speak.He was paid $2000.

Early in the 19th century, Hiram Moore of Climax, Michigan, had invented a large combined harvester and thresher and used it on his farm. This "combine" cut, separated, and sacked the grain in one sweep. Moore was granted a patent on his combine in June of 1830. Together with his neighbor, John Hascall, Moore tested his invention, and succeeded in reaping and threshing 1,100 bushels of wheat in one day. Moore then built four more such machines and used them for twenty-two years. Moore also invented a grain cleaner. After his death, his descendants shipped his combine to California, where rain seldom interrupted harvesting procedures.[7]

Then, in 1847, McCormick built a factory in Chicago, which was mainly run by his brothers while he spent much time in Europe selling his invention. This led to his increased wealth. By 1870 he was a millionaire. McCormick understood business methods and knew how to advertise his product. Nevertheless, because of the cost of this machinery, there were some farmers who still walked behind a plow in the early twentieth century.

McCormick first called his company Cyrus H. McCormick and Brothers. He later changed the name to McCormick Harvesting Machine Co. That factory was destroyed by the great Chicago fire of 1871, whereupona new factory was built, leading to a merger between McCormick Harvesting Machine Co., the Deering Harvester Co., and three additional harvester producers, which finally formed International Harvester Co. in 1902. That company controlled 90% of the grain binder business. From then on IH produced a line of tractors, leading to powerful tractors and later to the manufacture of trucks. In 1985 the Tenneco Corporation bought International Harvester, later to merge with the Case

Corporation and finally to Navistar, which has 15,000 employees and earns more than $12 billion annually.[8]

Steam engines for driving the harvester machines came into prominence in the 1880's. Daniel Best of California built the first combine equipped with an engine which took steam from the boiler of the traction engine and sent the power to the threshing unit.[9]

In 1890, Holt Brothers invented a leveling device which permitted the combine to be used on side hills. Because much wheat grows on hills in Washington State, this invention led to an increase in wheat production in Washington from 6 million bushels in 1890 to twenty million bushels in 1920.[10]

Another agricultural device developed in California was the vehicle stand rice dryer, which allowed taking rice directly from the field and drying it from a moisture content of 20% down to 14%, so that the milling yield of such rice became exceptionally high. This device had been invented in Italy and was then adopted in California.

When hand methods of rice production are compared to the mechanized method of rice production, it becomes evident that hand methods require 900 hours of work per acre while mechanical methods reduce this to 7.5 hours per acre.[11]

In 1929, the crusher mower was invented by E.B. Cushman in California. This device crushed the stems of alfalfa, which hastened drying 35% to 50% faster than hay cut with a standard mower.[12]

Multiple bean combines were developed in California in the 1920's. The bulk of the threshing was done by a slow speed cylinder. Together with some later improvements, this method decreased the damage to the seeds. After California farmers introduced high yielding flax into that state, this too was successfully harvested by using one steel and one rubber covered cylinder in the process.[13]

A labor shortage provoked by the Second World War led to a spectacular labor saving invention in the sugar beet industry.Using a long handled hoe required ten man hours per acre and twenty man hours per acre for whole seed planting. A machine developed in 1941 led to the result that, in 1944, over 900,000 acres could be planted, with savings of about 9 million man hours.[14]

Mechanical sugar beet harvesters were first used in 1943. Before that, special plows were used and then loaded into trucks by hand. Then John Powers invented a beet harvester which seized beets by their top according to the diameter of the beets. Lloyd Schmidt invented a beet harvester which spiked the beets after removing them from hard soil. By 1945 there were 3,695 machines in use harvesting beets.[15]

Mechanization of cotton picking has been attempted for over a century and succeeded in the 1920's in Texas and California. By 1949, only ten percent of cotton was picked by machines, but by 1962 this had risen to ninety percent. Machine picking also saved a great deal of money, as machine picking cost $25.76 per bale while hand picking cost $45.00 per bale.[16]

Innumerable additional devices have been invented to increase American harvests and lower the cost of agricultural production.

III.

One of the most important consequences of McCormick's, Moore's and other inventions was the reduction in the number and proportion of farmers in this country, accompanied by a reduction of the number of farms and the increase in the size of farms.[17]

In 1900, forty-one percent of the American workforce was employed in agriculture. Over the course of the 20th century, the agricultural segment of the workforce declined rapidly. In 1930 the agricultural workforce was 21.5 percent of the workforce, and in 1945 it reached sixteen percent. It then declined to four percent in 1970, and in 2002 was only 1.9 percent of the total workforce. In 2011, only 3 million, or 1 percent of Americans, lived on farms. Of these, 751,000 worked as farmers, which represented 0.5% of all employed Americans, although the rural American population was 51 million or 16.5% of the total American population. The U.S. Census Bureau defines as "rural" small towns of 2,500 people or less outside of urban clusters. These changes came about not only because of the mechanization of agriculture but also because of out-migration and a reduction of fertility among all Americans.

The numerous people living in rural areas who do not farm are generally commuters who work in nearby cities or in factories or businesses which have relocated in rural areas outside city limits.[18]

In the early 21st century, nearly 79% of American farms may be called "small scale." These farms are in the hands of farm families and average gross sales of $50,000 annually. These farms produce only 6% of sales of farm products, while the top 6% of corporate farms, with sales of over $500,000 annually, produce 62% of all sales. In 2010 there were 2,200,210 farms in the United States. Of these, 1,230,800, or 56%, sold between $1,000 and $10,000 worth of products per year, another 27%, or 594,850 farms, sold between $10,000 and a $100,000 of products in 2010, and another 7%, or 149,050 farms sold between $100,000 and $250, 000 of products in 2010. Only 99,510 farms or 4.5% of all farms sold between $250,000 and $499,999 worth of products that year and 6% of farms sold more that $500,000 worth of products in 2010. 1.5% of corporate farms reached a nearly $1 million income. The largest corporate farmers own 445,520 acres of farmable land, or 48% of the total American farm land of 919,990 acres.[19]

Because corporate farming is viewed by many private farmers as a threat to their lifestyle and income, nine states have prohibited corporate farming. These states are Iowa, Kansas, Minnesota, Missouri, Nebraska, North Dakota, Oklahoma, South Dakota, and Wisconsin. Corporations who own large tracts of land have attempted to defeat these laws by invoking the interstate commerce clause of the U.S. Constitution, which is in Article I, Section 8, Clause 3.

According to that clause, Congress has the power "to regulate commerce with foreign nations, and among the several states and with the Indian tribes."[20]

The dispute concerning the state laws prohibiting corporate farming centers on the issue of whether or not corporate, industrial farming is detrimental to the communities in which it is practiced. This issue has been the object of research for many years, beginning with the work of Walter Goldschmidt in the 1940's.

Goldschnidt was an anthropologist at the University of California. He found that corporate farming led communities to have a smaller middle class than similar communities without corporate farming. He further discovered that there were more "hired hands" in corporate farming communities than in communities which relied solely on individual farming. Communities with large corporate farms also produced lower family incomes and more poverty.[21]

There is considerable evidence that "industrialized" or corporate farming influences the number of jobs available, the quality of life, and the political consequences, in that non-resident, outside owners/managers of corporate farms directly influence the local governments to the detriment of all other citizens. Likewise, crime, family instability, high school drop-out rates, health, and the need for social services are all affected by the intrusion of large scale farm corporations into the community life of rural towns and villages.[22]

Large industrial farms also affect the old and the poor, as rents and real estate rise in cost when large corporations bring in highly paid associates who drive these expenses up. These problems may not be visible immediately, but become entrenched in affected communities decades after the arrival of corporate farming. When local farmers go to court to prevent corporations from acquiring large tracts of land, the courts generally base their decisions on the studies made by sociologists and others concerning the quality of life affected by industrialized farming.[23]

Industrial farms also shrink the population relative to family farms and create greater social stratification, as hired laborers and college educated management move into a community. This in turn leads to increased psychological stress and a deterioration of neighborly relations.[24]

Although large scale farming became successful only after the invention of mechanical agricultural devices, there have been a number of earlier efforts at creating corporate farms. These efforts have been made for over three hundred years, with no success until the 1950's. One example was the Puget's Sound Agricultural Company, which sought to supply wheat, livestock, and dairy products to the Russian fur traders in Alaska. That company lost money and finally ended its operations in 1870, as more and more small farmers raised the cost of labor.[25]

It wasn't until 1968 that the U.S. Department of Agriculture first reported on the extent of corporate farming in this country. At that time the department estimated that there were then 13,313 corporate farms in the U.S.A. These farms comprised 1% of all commercial farms then in operation and about 7 percent of all farmland, accounting for 8 percent of gross sales of farm products.[26]

The United States has 2.3 billion acres of land. 375 million acres are in Alaska and cannot be used for agricultural production. The land area of the "lower 48" is approximately 1.9 billion acres. Of this acreage, California alone has 103 million acres, Montana has 94 million acres, Oregon has 60 million acres, and Maine has 20 million acres, for a total of 277 million acres in just four states. About 349 million acres in the U.S.A. are planted for crops. Wheat is farmed on 62 million acres and is expected to be used for human consumption. Most of the other acreage is used to feed livestock. Much of the remainder is forest land.[27]

The U.S. Census issues a Census of Agriculture. This census reported that in 1986, corporation farms held 80,831,000 acres. Included in this immense acreage is the California farm of J.G. Boswell, which alone includes 200,000 acres and which demonstrates that large farms may be in private hands.

Large farms were at one time dependent on a large supply of labor which was seldom available, particularly as long as there was an American frontier to which so many immigrants and natives moved continuously. It was only after the closing of the frontier in California in about 1890 that labor became more available. By then, however, the combine had come into existence, leading to the experiments in the Great Plains in the 1920's.

In the 1950's, land prices were relatively low and Congress was continuing farm subsidies which made investment in farm machinery and in land attractive and led to the development of more corporate farming. This led to the belief that size matters. Yet, research has discovered that large farms are not necessarily an advantage, because uncertainty and the cost of coordination reduce the advantages size might bring. Furthermore, the cost of coordinating farm production can outweigh its advantages.[28]

Size does have some advantages in the farming communities. Large farms can exercise market power, both in buying and selling, as large farms gain volume discounts not available to small enterprises and also access favored tax treatment. The reason for this difference lies in the fact that smaller farms seldom have enough income to benefit from any tax concessions. In addition, it is evident that farm technology is expensive and therefore much more accessible to large scale corporate farm operations than to small farming enterprises. Nevertheless, the majority of farms in the United States as of 2012 are still in the hands of family farmers, however tentative this may be.[29]

IV.

The agricultural revolution which began with the introduction of the McCormick reaper was further enhanced by the use of science as well as technology in promoting the increase in the food supply in the United States, while also reducing the labor needed to achieve that.

There are a number of scientific methods now in use on American farms which came about as the more primitive farming methods were abandoned and more profitable scientific methods evolved. Included in these methods were the animal sciences, crop production, soil science, genetics, food science, horticulture, and an understanding of biology and chemistry.

Animal husbandry is commonly defined as an aspect of agriculture dealing with the care and breeding of domestic animals such as cattle, goats, sheep, hogs, and horses. The domestication of animals was of course one of the most important achievements in the prehistoric transition of humanity from hunting and gathering to agriculture. It has been estimated that the sheep were first domesticated around 9000 B.C.E. in the vicinity of the Tigris and Euphrates rivers in the area now called Iraq. By 6500 B.C.E., domestic goats were kept there, followed by the pig in 5900 B.C.E. The horse was first domesticated in the Eurasian steppes. Livestock was already known to the Romans. In modern times, veterinary medicine has improved the health of animals, decreased their abuse, and has contributed immensely to the food supply as well as making the raising of animals cost effective for farmers.[30]

Students who study in this area learn the origin of food, learn which breeds of animals are most productive, and study animal behavior. In addition, students in these college departments of animal science learn how to identify animal diseases such as spongiform encephalopathy or mad cow disease. The curriculum at the leading schools of animal science such as Iowa State University, Texas State University, and others includes biology, organic chemistry, animal physiology, cell biology, animal diseases, farm management, animal behavior, animal nutrition, and genetic engineering.[31]

Genetic engineering refers to technologies that are being used to change the genetic makeup of cells. Genes are chemicals which determine an organism's traits. Genetic engineering allows scientists to move genes across species, thereby creating new organisms, because the traits of one organism have now been transferred to another organism. This means that farmers can now escape the restrictions of nature when breeding animals. For example, a farmer may mate a black and white cow to a brown cow and produce a calf of a different color through sexual reproduction. Genetic engineering is purely technical and not dependent on sex. Therefore, genetic engineers can create foods not found in nature.[32]

The most famous experiment in modern biology was engineered by Dr. Ian Wilmut when, in 1997, he "cloned" Dolly the sheep (clone is Greek for twig). This dramatic success led to the development of the embryonic stem cell research field and the unfounded fear that humans would be "cloned" next. Frogs had been" cloned" but the frogs did not live past the tadpole stage.

There is good reason to believe that genetic engineering can eventually lead to the rehabilitation of patients with spinal chord injuries and other disabilities. However, political considerations have deprived scientists of the funds needed to conduct research into these possibilities.[33]

The science of horticulture, (hortus=garden) is undoubtedly as ancient as the "hanging gardens of Babylon" built by the king Nebuchadnezzar around 600 B.C. and considered one of the seven wonders of the ancient world. In America, horticulture influenced the development of the country in economics, social development and even international relations. Today, horticulture is called biotechnology and is characterized by four major "themes." The first of these is the introduction of plants brought to America by the colonialists of the 17th century, most of which failed to develop in the New World. The first reason for this failure was that Americans of the 17th century and beyond did not have the resources available to wealthy English landowners needed to propagate their plants. Moreover, there were American insects which did not exist in Europe which ruined many efforts at developing horticulture in this country.[34]

Horticulture has developed numerous areas of study. Arboriculture is the study of trees, shrubs, vines, and perennial plants, floriculture deals with the production of flowers, landscape horticulture deals with the production of landscaping plants, and olericulture deals with the production of vegetables. The production of fruits is called pomology, and viticulture is the production of grapes.

Those who wish to become horticulturists study courses in biology, botany, entomology, chemistry, genetics, physiology, and statistics.

Soil science is another important contribution to the promotion of scientific agriculture. This refers to the study of soil as a natural resource on the surface of the earth and includes the study of soil formation, classification and mapping, fertility of soil, and soil management. Soil science is complex and includes at least eleven specialties. Soil science is concerned with the use of food producing land in view of an ever growing population, as well as the use of water resources and land degradation.[35]

The consequences of these developments in soil science are that American farmers have increased food production from feeding four others in colonial times to feeding 130 others in the 21st century. Meanwhile, the population of the United States has increased from 76 million in 1900 to over 310 million in 2012. The average acreage of farms also increased considerably since 1920. Then, the average acreage farmed was 157 acres and the farm population represented 21% of the labor force in the United States. In 1940, the average acreage of American farms was 175, as the farm labor force had declined to eighteen percent. In 1960 the farm labor force declined more, to 8.3% of the labor force, as the average acreage farmed increased to 303. In 1970 the average acreage farmed by 4.6% of the American labor force rose to 390, and in 1980 the average farm acreage had risen to 426, as 3.4% of the labor force worked on farms. During the first decade of the 21st century, the average farm size in the United States rose once more, to 449 acres, as the farm population declined to 1.8% of the labor force.[36]

Since the principal reason for agriculture is the production of food, food science is another area of concern for those involved in farming. Food science deals with human and animal nutrition through the use of chemistry.

Food science also involves the measurement of food ingredients and microbiological food safety with respect to illness caused by food by means of detecting food borne pathogens. This leads food scientists to the study of molecular biology and biochemistry. Plant and animal husbandry and environmental studies augment these concerns.[37]

The United States has twenty-seven colleges of agriculture. The majority of these schools were founded as a consequence of the Morrill Act of July 2, 1862.That law, signed by President Abraham Lincoln, allotted 30,000 acres to each senator and representative. This legislation made it possible for each state to receive federal funds to establish a state college or university. These institutions then taught military tactics but also agriculture and were called "land grant" colleges because they were located on land granted each state. One example of a land grant college of agriculture is the Cornell University College of Agriculture and Life Sciences. This had been established in 1865 as New York's land grant institution. It changed its name several times and today includes 27 departments teaching 1. animal science, 2. biochemistry, 3. biological engineering, 4. biological sciences, 5. biometry, 6. soil sciences, 7. food sciences, 8. horticulture, 9. international agriculture, 10. landscape agriculture, 11. microbiology, 12. molecular biology and genetics, 13. neurobiology, 14. nutritional sciences, 15. plant biology, 16. earth and atmospheric sciences, 17. ecology, 18. entomology, 19. plant breeding and 20. plant pathology.A number of departments of related interest are also part of the curriculum. These include 21. biology and society, 22. development sociology, 23. informational science, 24. natural resources, 25. applied economics and management, 26. educationand 27. communication.[38]

An example of a lesser known college teaching agriculture is South Dakota State University at Brookings.Here students are also taught biosystems engineering, biology and microbiology, dairy sciences, economics, horticulture, and forestry and veterinary and biomedical sciences. Each department has a number of courses leading to degrees in these fields as well as pre-professional programs in medicine, dentistry, optometry, occupational therapy, veterinary science and others.[39]

V.

The invention of farm machinery and the use of fertilizers gradually led to overproduction of farm products and therefore depressed prices. Added to this malaise was the depression of 1929-1939, which struck first and hardest at the nation's farmers. This meant that farm income in 1932 was less than one third of what it had been in 1929, as farm prices fell fifty percent while goods and services farmers had to buy declined only 32 percent. Several efforts had been made by Congress to deal with the farm problem as early as 1920, although the first failed measure undertaken by the government was the Agricultural Marketing Act of 1929.This act was followed by the Agricultural Adjustment Act of 1933, which sought to restore farm purchasing power to the 1909-1914 level. Later, Congress passed the Jones-Costigan Act, which required minimum wage payments to field workers. However, in 1936, the U.S. Supreme Court declared the AgriculturalAdjustment Act unconstitutional.[40]

Subsequently, the administration of Franklin Roosevelt used a number of other strategies to help farmers, including the "parity" program, which was defined as the ratio of purchasing power of the net income per person on a farm to that of income of persons not on a farm during 1909-1914. These measures resulted in government ownership of numerous agricultural products, which were then used to feed the armed services during the 2nd World War. The fact is that none of the "New Deal" measures to alleviate the misery of the depression really worked. In fact, many of these measures made the situation worse. It was the Second World War which ended the depression, not the government measures.

After the Second World War, the farm programs were continued, and are now more than seven decades old. This was the result of the Agricultural Act of 1949, which is in effect to this day, although repeatedly amended. According to this legislation, sixty million acres were idled by 1960, and by 1980 nearly 20 percent of U.S. farmland was idled. This was meant to keep prices up as less was produced. Then in 1985 Congress passed The Food Security Act. This led to lower price support and reduced the accumulation of stocks and released land back to production. When in the 1990's high prices led to a considerable decline in price support payments to farmers, farm organizations demanded that these payments become fixed payments, no longer linked to prices of crops. These demands led Congress to pass the Federal Agriculture Improvement and Reform Act of 1996, which replaced deficiency payments with contract payments, so that farmers now got payments even when prices were high. This law set payments in advance for several years, so that farmers knew they would get payments even when prices were high. The cost to the taxpayers jumped from $4.6 billion in 1996 to $32.2 Billion in 2000.[41]

These programs have been part of farm income for so long that farmers can hardly imagine having to do without them. The agricultural community is wedded to this money, and politicians who need constant reelection will not eliminate them in favor of a free market. At the end of the first decade of the 21st century, most farm program payments go to a small number of the wealthiest farm owners, some of whom were paid $750,000 a year.[42]

In May of 2011, the House of Representatives Appropriations Committee voted to lower that threshold to $250,000. Unless this amendment is passed, the government, i.e. the taxpayer, would continue to spend $5 billion a year on subsidies for farmers whose income is far above that of the average American working man.[43]

In 2008, Congress passed a farm bill containing 15 Titles. Title I covers support for commodity crops, which deals with income support of growers for some commodities, such as cotton, rice, peanuts, sugar, and dairy. Title II deals with conservation, land retirement, and resource protection. Title III includes agricultural trade and food aid. This allows for an international food aid program as well as exports related to World Trade Organization programs. Title IV, concerning nutrition, includes the food stamps program and other nutrition assistance. Title V, concerning farm credit, refers to federal direct payments and guaranteed loan maintenance. Title VI, concerning rural development, seeks to support rural broadband access as well as coordination with local, state, and other federal programs. Title VII deals with agriculture research and has to do with the biosecurity program and biotechnology organic production. Title VIII includes forestry management and agro forestry programs. Title IX, concerning energy, supports the purchasing of renewable energy systems, together with an educational program. Title X covers the horticulture programs and covers fruits, vegetables, and organic agriculture. Title XI deals with the livestock program and the country of origin labeling requirements and meat and poultry state inspection. Title XII provides crop insurance and disaster assistance. Title XIII deals with commodity futures and the reauthorization of the Commodity Futures 2nd trade Commission. Title XIV is labeled "miscellaneous" and includes provisions to assist limited resources farmers and also provides for agricultural security. Trade and Tax Provisions are covered by Article XV, as is disaster assistance. That farm bill expires at the end of four years. Therefore a congressional committee held hearings in 2011 concerning the next farm bill to be considered in 2012.[44]

On May 31, 2011, the U.S. House of Representatives Committee on Appropriations approved an agricultural appropriations bill including 125.5 billion in both discretionary and mandatory funding. The bill reduces discretionary spending by $2.7 billion from the 2011 level.[45]

VI.

On January 1, 1994, the United States became a party to the North American Free Trade Agreement, which includes Canada and Mexico. This meant that all non-tariff barriers to agricultural trade between these three countries were eliminated. In addition, many tariffs were eliminated at once, while others were phased out over five to fifteen years. The United States and Canada already had a free trade agreement since 1989, removing almost all tariffs concerning agricultural trade.

As a result of these agreements, net farm income for 2011 rose by $15.7 billion or 19.8% since 2010, as it reached $97.4 billion. This is the second highest increase in farm income in 35 years and has been credited to the NAFTA. These increases are of course possible only because of the scientific revolution which American agriculture has achieved over more than a century. As a result, the Foreign Agriculture Service of the Department of Agriculture projects a decrease in government payments to farmers of 12.7 percent.

The North American Free Trade Agreement let to an increase in agricultural exports by American farmers to Canada and Mexico. In 1990, the value of agricultural products exported to Canada was $4,214,000,000. This rose to $16,253 million in 2008. The export to Mexico in 1990 was worth $2,560,000,000 but reached $16,025 million in 2008. Throughout the world, exports also rose considerably, as is visible by comparing the value of agricultural exports from 1990 through 2009 and beyond. In 1990, the value of agricultural exports worldwide was $39,495,000,000. This rose to $115,278 million by 2008 and is expected by the Department of Agriculture to reach $126.5 billion in 2011.[46]

All this was possible only because the United States farm community had become entirely mechanized and was using scientific agriculture to increase production. The impact of these technological and scientific methods on corn production are particularly important in Mexico, the land where corn was first produced and which is still farming corn with rather primitive methods.The import of American corn into Mexico has lowered the price of corn in Mexico, leading to hardship for Mexican farmers using outmoded techniques on small farms.[47]

In 1995, the United States joined the World Trade Organization, with far reaching impacts on world agriculture. Because American cotton growers are subsidized, other cotton growers, like Brazil, complain that they cannot compete with Americans in the world market. In addition to subsidies, American farmers also have the advantage of scientific developments not heretofore known. This refers to cellular biology, which has succeeded in allowing the transfer of genes from one plant or animal to another. Therefore, most of the cotton, soybeans, and corn grown in the United States has been genetically modified to resist some herbicides or make plants unpalatable to insects. This means that United States producers use genetically modified seeds on 109 million acres. Europeans generally do not use genetically modified seeds because they believe the food grown from these seeds is unhealthy for consumers.[48]

Ever since colonial days, American farmers have depended on exports as an outlet for their surplus production. Exports lagged only during the Great Depression of the 1930's. Thereafter U.S. farm exports steadily increased. Examples are the years 1997 to 2005 and later. In 1997, farm exports amounted to $57.3 billion, rising to $58.5 in 1998, $59.7 in 1999, $62.6 in 2000, $66.2 in 2001, $69.4 in 2002, $72.6 in 2003, $76.0 in 2004, $79.8 in 2005 and reached a record $137 billion in 2011 as American companies expand shipping capacities to meet the rising demand for American agricultural products around the world. This need for more food is mainly the outcome of the ever increasing world population, which, according to the U.S. Census Bureau, had reached more than 6.9 billion in June of 2011. The total export of American farm products increased by 26% between 2010 and 2011. Undoubtedly, this increase in American exports is also related to the decline in the value of the dollar, making exports cheaper for overseas customers.Real income in heretofore underdeveloped countries is also rising, so that people in poor countries will be able to spend more money on food and thereby benefit American agriculture. This can be true even if such giant populations as those of China or India undergo relatively small increases, which in sheer numbers are considerable.[49]

It is noteworthy that although there has been a slowdown in American industrial productivity during the years 2008 to 2011, there was no such slowdown in agriculture. As prices for agricultural products are reduced, world populations will buy more American food because they do not have the technological and scientific means of producing more agricultural products themselves.

The most important customer for American farm products in 2010 was China, which is the world's largest importer of cotton and soybeans. China also imported 600,000tons of US corn between April and May 2010. Japan and Mexico are the second and third largest importers of American farm products.[50]

The increase in exports of American farm products created a trade surplus of $47.5 billion in 2010, leading to the support of more than 1.1 million jobs nationwide. In 2012 Congress will consider several new trade agreements with Columbia, Panama, and South Korea, which will increase job opportunities even further. These countries are of course also engaged in making trade agreements with countries other than the United States.[51]It is evident that U.S. agriculture has made immense gains from its ability to export products around the world.

VII.

The agricultural revolution which began in the middle of the 19th century has had a profound effect on the lives of Americans. This is visible by inspecting the life expectancy tables since 1850. These show an ever increasing length of life since then, so that children born in 2010 are expected to live twice as long as children born in 1850.

Although longevity is related to inherited characteristics, there can be little doubt that the increase in diverse food production was a major factor in this development. In 1850, American males could expect to live 38.3 years. In 1900 this had increased to 48.23 years, and reached 67.55 years in 1950. In 2000, life expectancy for males born then was 74.8 years, and by 2010 this had increased once more to 78.3 years, with estimates for further increases in life expectancy in the future.

Life expectancy for American women has been longer than for males since 1850. In that year American females had a life expectancy of 40.5 years. That increased to 51.08 years in 1900, and further increased to 72.03 years in 1950. In 2000, American females born that year could expect to live 80 years, and by 2010, American females increased their life expectancy yet again to 81 years.[52]

Food production is not the only reason for these increases in life expectancy. Nevertheless, it is evident that the availability of diverse food as relatively low prices has promoted the health and welfare of millions of Americans during the last 150 years. Today, according to the Centers for Disease Control, as food is readily available in food stores and restaurants, many Americans are overweight and are therefore risking a shorter life than normal weight would allow.

Another consequence of the agricultural revolution has been the decrease in the age of menarche among American women. As meat and other proteins became more available as the size of Americans increased and the age of reproductive capacity decreased. The onset of menarche is influenced by body build, body fat, height-weight ratio, skeletal maturation, protein intake, and biological family inheritance.[53]

During the first half of the nineteenth century and later, menarche occurred at age 17 or 16. It then declined at a rate of three months per decade. This came about because nutrition and the general standard of living in the United States improved in the last 150 to 200 years. In earlier years, a lack of essential nutrients failed to trigger hormone synthesis essential for the development of menarche.[54]

In 2001, the mean age for the onset of menstruation in the United States had been reduced to 13.9 or 14 years. There were then differences in the onset of menstruation and social class. Those with more money and more education who ate more meat and expensive food experienced menarche at 13.5 years, while girls from poor immigrant families experienced the onset of menstruation at age 14.5. Clearly a distinction based on nutrition related to income.[55]

Now, at the second decade of the 21st century, the increase in the food supply at relatively affordable cost has led to a considerable increase in the average weight of Americans. That too is a product of the agricultural revolution. In 2008, the U.S. Department of Health and Human Services published that over two thirds of adults in the United States were overweight or obese. Obesity is a condition of exceptional overweight. Accordingly, 68 percent of men and 64 percent of women weigh too much. Of these, one third are obese. Therefore less than one third of American adults are at a healthy weight. The prevalence of overweight has steadily increased by over 13 percent since 1960.[56]

As nutrition improved, the height of Americans also increased. This contributed to the earlier onset of menarche. The American Journal of Epidemiology published a study in 2000 which showed that "girls who consumed more energy adjusted animal protein and less vegetable protein at ages 3-5 years had earlier menarche and girls aged 6-8 years with higher animal protein intake became adolescents with earlier peak growth."[57]

Height of Americans has increased over the past century. In 2009, the Centers for Disease Control reported that the average height of American men was 69.4 inches, or more than 5'9," and the average weight of American men was 194.7 pounds. The average height of American women that year was 63.8 inches or 5'4," and the average weight was 164.7 pounds.[58]

At the start of the 17th century, English men and women, including those living in New England, had average heights of 5'6" and 5 feet respectively. This remained the same until Americans reached heights of 5'7" and women 5'3" before the second half of the twentieth century.[59]

As the height of Americans increased and the menses occurred earlier, the age at marriage among Americans has risen. At the beginning of the 20th century, the average age at marriage for women was 21 and for men 24. In 2011, that had risen to 25 years for women and 27 years for men. Therefore, the years from arrival at the menses to marriage increased from a span of four years, i.e. 17 to 21 before the second half of the 20th century to a span of twelve years, i.e., from age 13 to 25, after the second half of the 20th century. Consequently, the agricultural revolution, the increase in the food supply, and the improvement of nutrition are responsible for the ever growing number of children born to single mothers. This is an important finding, as it needs to be juxtaposed with the oft repeated view that unmarried motherhood is simply a moral issue.[60]

The increase in the rate of births to unmarried mothers in the United States is indeed dramatic. Between 1930 and 1934, six percent of white women and twenty-five percent of black women were unmarried when giving birth to their first child.[61]

In 1970, the percent of births for unmarried mothers remained about 6% for white women, but had risen to 37.5% for black women. Thereafter a sharp rise in births to unmarried mothers was recorded. For white women, the percent of live births to unmarried mothers was 7.1% in 1975, 11.2% in 1980, 14.7% in 1985, 20.4% in 1990, 25.3% in 1995, 27.1% in 2000, 30.5% in 2004, 31.7% in 2005, 33.3% in 2006, 34.8% in 2007, and reached 38.0% in 2010. When the black rate of births to unmarried mothers is included, the overall percent of unmarried motherhood in the United States was 41% in 2010.[62]

Although there are numerous reasons for the increase in births to unmarried mothers, it is evident that these increases could not have occurred before the decline in the age of menses among American women ... and that decline is the product of improved nutrition and the industrialization of American agriculture.

VIII.

The future of American agriculture appears to be substantially different from its past. Prior to the 21st century, farming in America was always plagued by chronic surpluses, which reduced the income of American farmers. In the 21st century, worldwide demand for agricultural products has caught up with supplies, so that overproduction can hardly be expected again. That is why the most important farm policy issue in the past was the support of prices in view of so much production. That is why farm income before 1960 usually amounted to only 51% of non-farm income. Food prices were generally low in those years. In the first decade of the 21st century, food prices have steadily increased, although the proportion of income spent on food increased only from 9.7% to 9.9% between 2004 and 2010. Twenty years earlier, 11.7% of income was spent on food in America. It is also noteworthy that 48.5% of money spent on food is spent away from home, i.e. in restaurants. Fifty years ago, only 25% of food money was spent away from home.[63]

Americans may not know it, but the truth is that American agriculture has made the U.S.A. the envy of the world due to our ability to use science and technology to benefit everyone, as everyone must eat every day.

Summary

For centuries, before the mechanization of agriculture, food production was an endless task. Therefore, the invention of the reaper by Cyrus McCormick and Obed Hussey was a revolutionary event in the history of mankind. Consequently, the combine and other labor saving devices made it possible for fewer and fewer farmers to produce more and more food in less time than ever imagined in earlier years. Furthermore, scientific farming led to an immense oversupply of food products as well as the need for extensive education in managing agriculture. Higher education in agriculture became widespread, as government became heavily involved in subsidizing the agricultural enterprise. Moreover, better food led to greater longevity and other changes in the American population. In the 21st century, international trade agreements have created worldwide markets for American farmers.

Notes

1. Arthur Cecil Bining, *A History of the United States,* (New York: Charles Scribner's Sons, 1950) vol. 1, page 91.

2. Timothy Egan, *The Worst Hard Times* (Boston and New York: Houghton Mifflin Co., 2006), p. 47.

3. McCormick Collection, *McCormick Reaper Centennial Source Material,* (Chicago: State Historical Society of Wisconsin, 1931), pp. 23-28.

4. Edward T. O'Donnell, "Cyrus McCormick Invents the Reaper," *Irish Echo,* (June 19, 2002), p. 1.

5. Lee Grady, "McCormick Reapers at 100: Marketing the Machines that Revolutionized World Agriculture," *The Wisconsin Magazine of History,* vol. 84, no.3, (Spring 2001), pp. 20-21.

6. Thomson Gale, *World of Inventions,* (New York: Thomson Gale Corporation).

7. F.H. Higgins, "The Combine Parade," *The Farm Quarterly,* no. 4, (Summer, 1949), p. 42.

8. Tim Engstrom, "A Look at International Harvester History," *The Austin Daily Herald,* (May 17, 2011), p. 3.

9. F. Hal Higgins, "97 Years of Combining in California," *California Farmer,* vol. 25, (March 1950), pp. 280-281.

10. H.D. Witzel and B.F. Vogelaar, "Engineering the Hillside Combine," *Agricultural Engineering,* vol. 36, (August 1955), pp. 522-525.

11. Roy Bainer, "Harvesting and Drying Rough Rice in California," *Rice Journal,* vol. 47, (July 1944), pp. 12-14.

12. Roy Bainer, "Preliminary Trials of a New Type Mower," *Agricultural Engineering,* vol. 12, (May 1931), p. 165-166.

13. Paul F. Knowles, "Flax Production in the Imperial Valley," *California Agricultural Experiment Station Circular,* Davis, California No. 480, (July 1959).

14. Leonard J. Arrington, *Beet Sugar in the West,* (Seattle: The University of Washington Press, 1966), pp. 145-146.

15. John B. Powers, "The Development of a New Sugar Beet Harvester," *Agricultural Engineering,* vol. 29, (August 1948).

16. James R. Tavernetti and Lyle M. Carter, "Mechanization of Cotton Production," (Davis, CA: *California Agricultural Experiment Station,* Bulletin 804, 1983).

17. Grady, "McCormick's Reaper," pp. 13-20.

18. Allison Tarmann, "Fifty Years of Demographic Change in Rural America," (Washington, DC: *Population Reference Bureau,* 2009), p.1.

19. Paul Sueper, "Number of Farms and Farms by Sales Class," (Washington, DC: United States Department of Agriculture, National Agriculture Statistics Service, February 18, 2011), p. 1.

20. Harrison M. Pittman, "The Constitutionality of Corporate Farming Law in the Eighth Circuit," *The National Agricultural Law Center,* (June 2004).

21. Walter Goldschmidt, *"Large Scale Farming and the Rural Social Structure," Rural Sociology,* vol. 43, (1978), pp. 362-366.

22. Donald Boles and Gary Rupnow, "Local Governmental Functions Affected by the Growth of Corporate Agricultural Land Ownership," *Western Political Quarterly,* vol. 32, (1979), pp. 467-478.

23. David Gough and Don Elbourne, "Systematic Research Synthesis to Inform Policy, Practice and Democratic Debate," *Social Policy and Society,* no.1, (2002), pp. 225-236.

24. Douglas Jackson-Smith and Gilbert W. Gillespie, "Impact of Farm Structural Change on Farmers' Social Ties," *Society and Natural Resources,* vol. 18, (2005), pp. 215-240.

25. Michael L. Olsen, "Corporate Farming in the Northwest: the Puget's Sound Agricultural Company," *Idaho Yesterdays,* vol. XIV, (Spring 1970), pp. 18-23.

26. U.S. Department of Agriculture, *Corporations with Farming Operations,* Agricultural Economics Report No. 209, (Washington, DC: U.S. Government Printing Office, June 1971).

27. Ruben N. Lubowski, Marlow Vesterby, Shawn Bucholtz, Alba Baez, and Michael J. Roberts, "Major Uses of Land in the United States," (United States Department of Agriculture, Economic Research Service, 2002):4.

28. Hugh M. Drache, *The Day of the Bonanza,* (Fargo, ND: North Dakota Institute of Regional Studies, 1964).

29. Kenneth R. Krause and Leonard R. Kyle, "Economic Factors Underlying the Incidence of Large Farming Units," *American Journal of Agricultural Economics,* vol. 42, (December 1970), p. 753.

30. Louis N. Manger, *A History of the Life Sciences,* (New York: M. Dekker, Publishers, 2002) pp. 1-2.

31. No author, "What is an Animal Science Major?" *Degree Directory,* http://degreedirectory.org.

32. No author, "What is Genetic Engineering?" (Cambridge, MA: Union of Concerned Scientists, 2003), p. 1.

33. Sally Lehman, "Dolly's Creator Moves Away from Cloning and Embryonic Stem Cells," *Scientific American,* (August 2008):1.

34. Philip J. Pauly, *Fruits and Plains: The Horticultural Transformation of America,* (Cambridge, The Harvard University Press, 2008), p.9 and p. 33.

35. Eric C. Brevik, and Alfred Hartemink, "Early Soil Knowledge and the Birth and Development of Soil Science," *Catena,* vol. 83 (2010), pp. 23-33.

36. United States Government, Environmental Protection Agency, "Ag 1010" (Washington, DC: Government Printing Office, 2007), p.1-8.

37. *Journal of the Science of Food and Agriculture*, "Instructions to Authors."

38. http://cornell.edu, Cornell University College of Agriculture and Life Sciences.

39. http://sdstate.edu, South Dakota State University School of Agriculture and Biological Sciences

40. Douglas E. Bowes, Wayne D. Rasmussen, and Gladys L. Baker, "History of Agricultural Price Support and Adjustment Programs, 1933-1948," (Washington, DC: The U.S. Department of Agriculture, Information Bulletin 485, 1985), pp. 1-20.

41. Bruce L. Gardner, "Does the Economic Situation of U.S. Agriculture Justify Commodity Support Programs?" *AEI Agricultural Policy Series: The 2007 Farm Bill and Beyond.* (Washington DC: The American Enterprise Institute, 2007).

42. Daniel A. Sumner, "Targeting Farm Programs," *Contemporary Policy Issues,* vol. 9, (January 1991), pp. 93-106.

43. Mary Clare Jalonick, "Farm Subsidies: House Appropriations Committee Approves Cuts," *The Huffington Post,* (May 31, 2011), p. 1.

44. Renee Johnson and Jim Monke, "What Is the Farm Bill?" *Congressional Research Service Report for Congress,* (December 10, 2010), p. 2.

45. Jennifer Hing, "Appropriations Committee Approves Fiscal Year 2012 Agriculture Bill," The Capitol, Washington DC, "H 307 (May 31, 2011), p. 1.

46. U.S. Department of Agriculture, Economic Research Service and Foreign Agricultural Research Service "Foreign Agricultural Trade of the United States" (February 2010), Table 850. U.S. Census Bureau *Statistical Abstract of the United States.* (2011), p. 547.

47. Tim Weiner, "Manzanillo Journal: In Corn's Cradle, U.S. Imports Bury Family Farms," *New York Times,* (February 26, 2002), p. 4.

48. Clive James, "Global Status of Commercialized Transgenic Crops" *International Service for the Acquisition of Agri-biotech Applications,"* Brief 42, 2010.

49. No author, United States Department of Agriculture, "Ágricultural Exports and Net Income 1997-2005," USDA Economic Research Service, (February 2006).

50. Tom Vilsack, "US Farm Exports Set Record in 2011," (Washington, DC: The Department of Agriculture Economic Research Service, 2011).

51. Alan Bjerga, "U.S. Farm Exports May Increase to Record $137 Billion," *Bloomberg,* (May 26, 2010), p. 1.

52. *National Vital Statistics Reports,* vol. 54, no.19, (Department of Health and Human Services, National Center for Health Statistics, June 28, 2006). See also: *2008 National Population Projections,* U.S. Census Bureau, (August 2008).

53. Lana Thompson, "Menarche," In: *International Encyclopedia of Marriage and the Family,* (New York: Macmillan Publishers, 2003).

54. Ibid.

55. G.J. Engelmann, "The Age of First Menstruation in the United States," *Medical News,* (June 22, 1901), p. 1.

56. K. M. Flegel, M. D. Carroll, G. L. Ogden and L. R. Curtin, "Prevalence and Trend in Obesity Among US Adults" *Journal of the American Medical Association,* vol. 303, no. 3 (2010), p. 235-241.

57. Catherine S. Berkey, Jane D. Gardner, A. Lindsay Frazier and Graham A. Colditz, "Relation of Childhood Diet and Body Size to Menarche and Adolescent Growth in Girls," *American Journal of Epidemiology,* vol. 152, no.5, (2000), p. 446-452.

58. *Anthropometric Reference Data for Children and Adults: United States, 2003-2006,* Centers for Disease Control and Prevention, National Center for Health Statistics, (March 6, 2009) tables 4, 6, 10, 12, 19, 20.

59. David Dunning, "The Average Height of Humans Over Time," *Health,* (September 10, 2010), p. 1.

60. Tricia Ellis Christensen, "How has the average age at marriage changed over time?" *Wise Geek,* (March 26, 2011).

61. Amara Bachu, "Trends in Marital Status of U.S. Women at First Birth: 1930-1994," U.S. Bureau of the Census, Population Division, Working Paper No. 20 (March 1998).

62. Center for Disease Control, "Key Birth Statistics" (Atlanta, Ga. April 6, 2011), 1.

63. USDA Economic Research Service , "Americans Spend Less than 10% of Disposable Income on Food," (July 19, 2006), p. 1.

Chapter Four

The Sewing Machine, the Textile Industry, and the Role of Women

I.

The invention of the sewing machine and the washing machine, as well as the invention of the refrigerator, liberated women from household drudgery and permitted them to access a life of educational and professional achievement. Prior to these inventions, women were constantly enslaved by the demands of their families. Poor women were of course far more subject to the immense strain which domesticity imposed on them than was true of wealthy women. The burden of this enslavement was particularly hard on the many immigrant women who came to the United States during the great migration from Europe between 1890 and 1924. This account illustrates this dilemma: "She had to look not only after the house and her husband and children, but also after the boarders even during a period of advanced pregnancy. When the time of her parturition came, the worn-out woman in the throes of childbirth had to stifle her cries of pain in order not to wake up the boarders." Even into the 1920's and 1930's, women sewed, cleaned, and washed throughout the day and into the night without the aid of any labor saving devices.[1]

The work of sewing was continuous and unending before the sewing machine. In 1867, it took 20,620 stitches to make an average shirt by hand at the rate of 35 stitches a minute, so that a competent seamstress could make a shirt in ten to fourteen hours.[2] The labor of the nineteenth century seamstress was captured most succinctly by Thomas Hood in his poem "Song of the Shirt."In twelve stanzas, Hood describes the misery of those who worked "With fingers weary and worn, with eyelids weary and red, a woman sat in unwomanly rags, plying her needle and thread." The poem described the horror of poverty and the cruel exploitation of those too poor to avoid such terrible slavery. "Oh Men with Sisters dear! Oh men with mothers and wives! It is not the linen you are wearing out, but human creatures' lives! Stitch-stitch-stitch, in poverty, hunger and dirt, sewing at once with a double thread, a shroud as well as a shirt."[3]

The development of the sewing machine over several decades seemed at first to be a liberating device which finally relieved women of this never ending task. For wealthy women that was indeed the case. In the early 19th century and before, when sewing machines were not yet available, sewing was a perpetual female task from generation to generation. Furthermore, sewing became a social event, as women sewed in "sewing circles" in the company of their friends. Wealthy women sewed embroidery, which could take weeks and even months and was regarded as a means of personal expression. Sewing was also used to raise funds for religious organizations and political parties while also serving as a gendered form of the division of labor. Upper and middle class men did not sew.[4]

For the poor, the sewing machine became a new taskmaster, a slave driver, and an instrument of degradation, misery, and desperation. The first sewing machines were made in England by Thomas Saint in 1790 and in France by Barthelemy Thimonier in 1830. The machine invented by Saint remained only a drawing until it was manufactured in slightly altered fashion by William Newton Wilson in 1874. The French sewing machine came into immediate use by its inventor until his factory was burned down by workers fearful of losing employment to the machine.[5]

In 1848, Elias Howe made the first successful American sewing machine, although Walter Hunt of New York had created a sewing machine in 1832 which failed. Elias Howe succeeded, not because he was the sole inventor of the sewing machine, but because he brought together existing inventions and techniques. The inventions used by Howe had been developed in Germany and Austria, and consisted of the eye-pointed needle, a stitch device, and a feeding mechanism invented in France and England. The ideas were integrated into a commercially useful machine by several American contributors during the decade of the 1840s. Thereafter, A.B. Wilson improved on the Howe machine until Isaac M. Singer invented the foot treadle. Singer then coordinated all the other inventions and manufactured sewing machines on a large scale. Singer was particularly successful in advertising his machines. He not only advertised in magazines, but also displayed them at international exhibitions such as the Word's Fair in New York City in 1853 and the Centennial Exhibition in Philadelphia in 1876. Free demonstrations for potential customers were another form of advertisement used by Singer. Singer built factories all over the United States and hired salesmen to promote the machines. He also sold sewing machines in Germany, Japan, Russia, Mexico, India, Ireland, and South America. These machines were deemed to reduce labor in the family. Singer gradually dominated the sewing machine market, so that by 1900 they were selling one half a million machines a year. As Singer and others increased their sales, the machines became less expensive, which led to the development of clothing factories, so that during the 19th century the clothing industry shifted gradually from household production to large scale production. Such large scale production became possible because interchangeable parts were produced, leading to the manufacture of everything from umbrellas to boots.[6]

Although sewing machines were mainly the product of European inventiveness, it was the United States which made these machines commercially feasible, and the immigrant worker the most reliable source of labor to work on these machines. Because so many poor immigrants were always available to manufacturers, it was the United States which created the "sweatshop" conditions in the clothing manufacturing business. The commercial use of the sewing machine meant at first that many garments which had always been made by hand could now be made by a machine. Hand sewers earned only $3 a week in 1840. In 1860 machines had been substituted for human power, so that a manufacturer could substitute 400 machines for thousands of hand workers. The workers who remained earned $4.00 per week.[7]

At once, factories employed women who had experience with making clothes at home. It has been estimated that 70% to 80% of employees in the "needle trades" in the United States and Canada during the 19th and 20th century were women.[8]

These women were not only subjected to exploitation by the manufacturers and owners of clothing factories, they were also victimized by the trade unions, who sought for years to exclude women's wage labor as part of their efforts to improve the wages of men. Even when, after the beginning of the 20th century, women were partially organized, women were still paid lower wages than men, even as women were segregated according to union contracts. This discrimination targeting women was in part motivated by the fear that women would seize men's jobs and by the belief that woman's job was to stay at home and provide for her family.[9]

Men viewed women employees in the needle trades with hostility because the lower wages earned by women reduced the job opportunities of men. Women also suffered from chronic occupational diseases which limited they functions as homemakers and mothers. Because the pay earned by women in the needle trades was so low, many seamstresses added to their income by prostitution.[10]

As early as the 1820's, ready made clothing for men supplanted the home made custom work. Standardized sizes came to be the norm, leading to the destruction of the traditional garment manufacturing procedures. This led to factory manufacturing of clothing and the abandonment of the journeymen-employer relationship, which provided for room and board for the employees. Now, employers laid off employees during any slack season and paid them only by the piece during short-term contract periods. This was made possible by the use of the sewing machines.[11]

Prior to the invention of the sewing machine, most women had worked extensively with the needle, as sewing was an endless job. Now, manufacturers contracted out parts of the work of clothing manufactures to women working at home. Competing with male garment workers, these women were paid a minimum. Some manufacturers attached a workshop to their dry goods shops, leading finally to the sweat shops of the late 19th and early 20th centuries.[12]

In time, this led to a division of labor, so that outwork and subcontracting became the norm in the industry. This led to "sweating," defined as "the effort by master tailors to procure the utmost of labor from journeymen tailors for the smallest possible remuneration."[13]

Thos who labored at home were almost only women and girls. Family members sometimes worked with friends from the neighborhood who needed the money, thereby reducing the minimum incomes of all these workers even more. Most of these poor laborers were immigrants. Because European Jews had some experience in the "needle trades," Jews were the majority in these sweatshops. This was also true of contractors, who generally were as miserable as those they employed. Contractors sought to underbid other contractors, so that wages became lower and lower even as homeworkers undercut each other.

This became possible because the work of making shirts and other garments was more and more reduced to a division of labor, allowing even children to do one task over and over. Thus, women and children entered the garment trades until few men were left to make a living there. As the ready-made garment industry expanded, more and more clothing was produced in factories and less was available to those working at home.[14]

II.

As we have already seen, it was the invention of the sewing machine which allowed the ready-made clothing industry to expand into the factories of the late 19th and early 20th centuries. At first, not all work was done in factories. Some work was still allocated to home production. Most important was the cutting of the material, which had to be laid out on a table and done with great care, lest failure to achieve a correct "fit" spoiled the investment in the materials. In 1850, a "band knife" was invented which allowed the cutter to cut more than one garment at a time. This was further improved when in the 1870's "long knives" where invented, which allowed cutting 18 garments at once.[15]

The sewing machine also revolutionized the speed at which a garment could be produced. It has been estimated that it took more than 16 hours to make a coat by hand and only 2.5 hours by sewing machine. The sewing machine reduced the cost of clothes and was cheap enough to be installed at home. At the beginning of the 20th century, almost all clothing manufacture was located in private homes, as the industry rested on the seemingly endless pool of cheap female labor who were exploited in the most ruthless manner.[16]

The wages of garment workers at the beginning of the 20th century were so low that one hundred years later it seems almost impossible that anyone could have lived on such incomes. Some women earned only 75 cents weekly while some earned $3.00 per week, a sum considered better than adequate. Those women who were employed in the contractor's residence earned $1.25 to $3 per week and those who worked at home received between $1 and $2 a week, or less than $52 per year in view of holidays and seasonal layoffs.[17]

Compared to the earnings of other occupations at the time, these earnings were abominable. At the turn of the 20th century, a house painter earned $840.00 a year and bricklayers earned $1,240 annually. Then, the average yearly earnings in the building trades were $1,600. Prices were also low compared to the 21st century. On January 1, 1901, a pair of men's shoes sold for $3.50 and an overcoat cost $6.50. Dress fabric cost 5 cents per yard in 1901, when the New York Times cost 1 cent every day.[18]

The exploitation of women workers was brutal. If an alteration needed to be made on a garment delivered to the contractor by a home worker, the contractor fined the worker a sum equal to the time needed to make the change. The women who did the work paid the fine rather than take the garment home, since the time needed to travel home and back cost more than the fine. Employers also "sold" sewing machines to home workers and charged the cost against their miserable wages. A sewing machine cost about $55, resulting in the deduction of $2 a month from the earnings of the affected woman.[19]

Garment workers labored sixty hours a week and often had to work overtime without extra pay because some customer was in a hurry to get her dress or suit. This meant that many a laboring woman would eat her dinner in five minutes and work additional hours just to keep her job. At home, many worked 16 hours a day and additionally had to wait at factory doors to ask for work. Women would wait two hours at a factory door to get a minimum amount of work to be done at home.These conditions continued to plague the garment industry into the better part of the 20th century.[20]

Children, too, were brutally exploited in the garment and other industries in the early 20th century. Because time was money, children were used at the youngest ages. For example, five-year-olds were forced to pull out basting stitches (basting stitches are extra long stitches used to hold two pieces of garment temporarily together).By age ten, children could attach buttons and deliver clothing bundles. Children became the assistants to their mothers and fathers as the whole family worked in sweatshop conditions to barely stay alive.[21]

The needle trades also caused numerous health problems for the women working there. Eye strain, headaches, fainting, back pain, and spinal distortions came about from constantly bending over the sewing machines. These machines brought with them physical exhaustion, leg pain from the constant use of the treadle, and a 'tremble' caused by the vibration of the machines.[22]

In the winter, life was particularly difficult for the sweat workers in the clothing industry. During the winter months, manufacture slowed down and many workers had no income then. In addition, they had to pay for fuel to heat the miserable tenements they inhabited. Each winter these unfortunate laborers lived on the few dollars they saved in the warmer seasons until work resumed in April. This forced many women into prostitution so that "seamstress almost became a euphemism for prostitute."[23]

Housewives, as non-employed women were called, often needed to supplement their husbands' incomes by "taking in" sewing in addition to their housework or farming responsibilities. Few men earned enough in the nineteenth and twentieth century to support large families without the extra money earned by mothers and wives. This led many farm women to order clothes through mail order catalogues, which first became available when Montgomery Ward started a catalog business in 1870. In that year, the farm population of the United States was 18,373,000 out of a total population of 38,500,000. Fifty-three percent of the U.S. labor force was engaged in agriculture on 2,660,000 farms averaging 153 acres. The meager income farmers earned then made it necessary for wives to contribute financially in some manner.[24]

The belief that women should engage in domestic production in the form of sewing and laundry was reflected in the curriculum of Iowa State College of Agriculture and Mechanic Arts and other such colleges, and the Smith-Lever Act of 1914, which created extension services to provide education to rural populations.[25]

This need for an additional female income in most American families persists into the 21st century.The Bureau of Labor Statistics reported that in 2010 the median income of American men was $874 per week, while the median income for American women was $704, or more than 19% less than median male income.[26]

Then, and now, homework is one of the few means by which women with children can earn extra income. This was even more important before women achieved the higher level of education available to them in the 21st century. The need to earn money at home was particularly important to single women with children then and now. Because women had few skills other than housework and needle slavery, women begged for work from stingy employers. In large cities, particularly in New York, women and children could be seen dragging heavy bundles of clothing to and from manufacturers for pitiful wages. In view of these hideous conditions, American needle trade workers began to form unions so as to maintain work standards, which were constantly deteriorating in view of the competition for work caused mainly by the influx of poor immigrants into the major American cities.[27]

These unions were at first too weak to gain any concessions from clothing manufacturers. It was not until the early twentieth century that the Amalgamated Clothing Workers of America and the International Ladies Garment Workers Union achieved at least some success in maintaining some livable standards for their members.[28]

That success came slowly, as there were numerous obstacles facing the organizers of the early labor unions. Among these difficulties was that most of the workers labored at home and therefore did not know one another, even if they worked for the same employer. Furthermore, many garment workers were ambitious to become employers themselves and therefore would not support unions. Then there were ethnic differences, as Jewish, Italian, and other immigrants competed for needle work. Furthermore, strikes were almost impossible to sustain, as the immigrant workers were so destitute that a strike meant instant hunger and deprivation. Moreover, there were so many new immigrants willing to work for a pittance that strikers could easily be replaced by newcomers.[29]

Because garment workers labored in unsanitary conditions, including very close proximity to other laborers as well as near starvation, many garment workers carried diseases such as tuberculosis, diphtheria, scarlatina, and smallpox. This led to the infection of clothing as well as an effort of some citizens to improve the lot of the garment "slaves" in view of these health hazards.[30]

III.

In the nineteenth century and in the early 20th century, American women still spent a great deal of time constructing and mending garments from fabric, which was still cheaper than buying ready made clothes. As a result, women of every economic stratum were expected to sew all the time. This led many people to the belief that the sewing machine was the great liberator of American women.

Originally, the sewing machine was viewed as "The Queen of Inventions," as the New York Times called the sewing machine "the best boon to woman in the nineteenth century." The sewing machine appeared to be a great labor saver and emancipator of women. In the 1850's, a sewing machine cost $125.00, which was quire expensive considering that the average income per family was only $500.00 a year. This led Singer to create the installment plan, allowing women to take home a sewing machine for $5.00 down.[31]

The sewing machine also became a status symbol in the 19th century. This was so because the machine cost a great deal of money, so that the sewing machine in the living room proclaimed to visitors that the family had the resources to own one. The sewing machine was encased in cabinets of different styles so that the machine would fit into the style of furniture in the room. As sewing machines became cheaper by reason of mass production, more and more poor people owned a sewing machine, so that it lost its appeal as a status symbol for the middle and upper classes.[32]

That sewing machines were in common use at the beginning of the 20th century can be seen by the introduction of the sewing machine into the pages of women's magazines. Consequently, the sewing machine was now concealed and no longer displayed. Now the sewing machine was widely used but invisible. However, the most important reason for the loss of status of the sewing machine was the rise of the ready-to-wear industry. This meant that bought clothes were cheaper than homemade clothes, and the sewing machine was no longer viewed as a labor saving device.

Nevertheless, until about 1910, many American women still sewed the bulk of their clothes. This was true not only because of the cost of ready made clothes, but also because women were taught to believe that sewing was an integral part of fulfilling her role as the family caretaker. Sewing gave women a sense of worth, as work outside the home was not often available and because most middle class men viewed it as demeaning if their wife worked for wages outside the home.[33]

Sewing was associated with white, middle class morality and Victorian ideals of domesticity. Furthermore, factory made clothing for women constituted only 25% of all women's clothes in 1890, a condition wich continued into the 1920's.It was far cheaper for women to buy fabric and make their own clothes than to buy them or to have them custom made by a professional dressmaker, which only wealthy women could afford. It was only after the 1920's that ready made clothes finally became affordable to the middle class, whose women nevertheless continued to make clothes at home. This was particularly true of farm women, who were so isolated that they seldom had an opportunity to buy anything in any store unless they traveled some distance to the nearest town. In fact, a survey showed that in 1926 a majority, i.e. 95%, of rural women still made their own clothes.[34]

The reason for this effort was the cost of buying clothes in view of the abominable income earned by working people. For example, an envelope machine operator earned $6.00 per week in 1911 and a corset maker earned $10 a week that year. A skilled "fitter" working in a department store earned $12.00 a week any time before the First World War (1914-1918).[35]

The final transition from producer of clothes to consumer was made during the decade ending in 1920, the year in which women first voted in federal elections and in which the number of seamstresses declined by over 60 percent since 1860. In the next decade, the ratio of dressmakers to the female population fell even further, so that if finally reached 2% in 1930.

After 1920, the sewing machine was stigmatized. By then, ready made clothes were almost universal and women who sewed their own were viewed as antiquated and old fashioned. Homemade clothes were viewed as amateurish and visibly inferior to machine made clothes produced by "professionals." In fact, homemade clothes were viewed as a source of embarrassment, as such clothes had become a clear sign of a family's economic status.[36]

Of course, by 1920, the entire clothing industry had come under scrutiny as an exploiter of cheap immigrant labor. These immigrants were paid so little that they could not afford to buy the clothes they made. Moreover, the "Triangle Shirtwaist Factory fire" of 1911 had increased the membership of the International Ladies Garment Workers' Union considerably, as the deaths of 146 women in that fire led to improved factory safety standards.[37]

The deaths of these many women were primarily the responsibility of the factory owners, who had locked the doors to the stairwells and exits. The fire caused people to jump from the eighth, ninth, and tenth floors into the street below, where they died on impact, even as hundreds looked on in helpless terror. There was only one internal fire escape, which buckled under the weight of so many bodies. As a result, numerous workers jumped to their deaths before any fire engines arrived.[38]

There are numerous testimonials written by the semi-serfs who worked under hideous conditions in these garment factories where the sewing machine became the symbol of their misery and enslavement. One such testimony was written by Rose Cohen, a sweatshop worker who survived the Triangle fire.

Rose Cohen, at age 14, started work at 6 a.m. and worked until 7 p.m., and often until late at night, six days a week, for $3 a week. She climbed several stories up inside the factory and there was forced to sit between two sweating men sewing sleeve linings, a truly dastardly task. As she finished each coat, an ugly boss examined the stitches and then gave her more work. In the same shop there were four sewing machines and sixteen people were working. In view of the abject poverty of the girls who worked there, the workers envied one another, as each viewed the other as a competitor for a piece of bread. The male bosses also engaged in sexual harassment, as there were no laws to protect working girls against this and because the number of recent immigrants willing to take any work, however cheap and distasteful, was limitless.

At home, Rose Cohen lived in a small "tenement" apartment, sleeping in a cold room and eating only rice. Her horrifying story was multiplied by thousands who suffered as much in New York and Boston, in Philadelphia and Montreal, and all places where immigrants could be exploited by brutal bosses.[39]

Clara Lemlich, another victim of these inhumane conditions, describes how the bosses paid a few workers a salary but paid most for "piece work." According to her, "regular work" paid about $6 per week. The girls who worked in these hellholes started at 7 a.m and stayed until 8 p.m. with only one half hour for lunch. The sewing machines were illuminated by gaslight except for the front row near a window. According to Lemlich, the bosses screamed epithets at the girls and threatened their livelihoods each day. There were no dressing rooms for the workers. The shops were filthy and the air foul. The bosses deducted from the meager wages any damage that may have resulted from the work or even if the material was damaged in the first place.[40]

As the sewing machine became more and more the tyrant who enslaved helpless workers, it lost importance in the home, as ready made clothes became the rule for most Americans after the beginning of the 20th century and particularly after the First World War. Because of the horrible conditions under which garments were produced, the workers finally united and began to strike. This didn't happen until 1916, although seventy-nine years earlier, in 1835, a textile strike was attempted in Paterson, N.J.[41]

In 1926 a textile strike took place in Passaic, New Jersey, which was remembered fifty years later at a public symposium at William Paterson College. At the time, the strike dominated public attention for nearly a year. The strike was caused by a ten percent wage reduction in the textile mills at Passaic. The labor force consisted mostly of foreign born workers, including many women and children. It was at first led by a lawyer whose communist affiliation alienated the American Federation of Labor. Upon his withdrawal, the strike came to an end in 1927, although Passaic remained a non-union town.[42]

An understanding of the conditions facing textile and garment workers can be gleaned from the letters written to President Franklin D. Roosevelt after the 1934 textile strike failed to bring relief to the near slaves in that industry. These letters include such statements as: "The labor conditions in the ... mills are no less than slavery." The writers told Roosevelt that they worked ten hours with wages not sufficient to live on. Another letter dealt with the inability of the textile workers to visit doctors because they earned too little to pay for medical care. The letter writer tells the President that the employers pay as little as they like and work people inhuman hours. Another letter decries that woman laborers have no "rest stools" but must stand hour after hour. More letters discuss the utter depravity of the employers who had no regard for the physical suffering of the employees dependent on them.[43]

All of this demonstrates that by the 1930's the sewing machine had become the oppressor of millions whose livelihood depended on it, and that the sewing machine was responsible for a great deal of social change in America. It created ready-to-wear clothing, improvements to the carpeting industry, bookbinding, and shoe and hosiery manufacture, as well as upholstery and furniture construction. These industries came about because of the invention of the industrial sewing machine. The industrial machines made a zigzag stitch which was later adopted in home machines. These industrial machines are constructed to endure factory conditions over a long period of time. Originally hand operated, today these machines rely on computers and electronic devices. Home machines usually have casings and are made of light metals such as aluminum for easy moving.[44]

Industrial sewing machines are constructed with great precision. Each machine is designed to perform only one function, so that a series of machines are needed which in succession create a finished garment. The basic part of an industrial sewing machine is called a bit. This is a frame which houses the machine. In the 21st century, sewing machines are controlled by the use of computers, creating electronic machines. Design software is now used to produce designs of all types. This software allows the designer to rotate, shrink, enlarge, and select different colors and stitches, so that operators can use these machines as a means of exhibiting their personal creativity.[45]

In 2001, the Swedish Viking computerized sewing machine came on the market. It can be linked to the personal computer, making it possible to create stitch patterns and embroidery that heretofore could only be produced by industrial sewing machines. The computerized machines are also great time savers in that it takes only half the time to produce things. The computerized machines are loaded with memory caches and, in 2001, had floppy disk drives. These machines have created a good number of hobbyist sewers who enjoy the sewing machines, which have in effect become computers with needles and thread attached. These new machines cost between $600 and $7000 and can produce as many as 1,000 stitches a minute, compared to the average of 250 in the old machines.

These computerized machines operate like a photoshop. These machines come with 500 or more preloaded stitches. They can produce colors by means of a color touch screen, allowing sewing hobbyists to create complicated patterns. That means that these machines have eliminated the painstaking operations of conventional machines, particularly as the color touch screen has brought hobby sewers into the high tech age.[46]

IV.

Employment in the garment industry was about 225,000 in 1900 and rose to 824,000 by 1950. Thereafter, the number employed in that industry in the United States steadily declined, until in the 1960s there were only 300,000 factory jobs in the New York City garment district producing almost 90% of the clothing sold in the United States. Fifty years later, only 12,000 seamstresses are employed in making twenty percent of the clothes sold in this country. These clothes are generally classified as belonging to the "high couture" category, a nonsense word derived from the Latin (consuere) for "stitching."[47]

Those who worked in these near-slave conditions had gradually succeeded in acquiring at least some protection from the exploitation imposed by the owner-bosses of the past. That protection was needed, not only because of the excessive hours demanded of the workers who were mostly women, but also because the rather primitive bosses, owners, foremen, and supervisors sexually harassed the many female employees. In 1964, Congress passed the "U.S. Civil Rights Act" which established the "Equal Employment Opportunity Commission." According to that law, "unwelcome" sexual advances are prohibited. All of this was of course not the case in 1900 or for sixty years thereafter. Instead, those who needed the jobs available in the needle trades were the victims of sexual harassment and cruel sexual - urinary "jokes." Bosses ridiculed the single girls working for them and the male coworkers enjoyed the victimization of the helpless girls who had no other income and no protection. In addition to the stupid "humor" and sexual innuendo employed by employers, there were also those who made sexual demands on the workers. A gender hierarchy existed in these sweatshops, with those on the bottom suffering the most.[48]

Sexual harassment made life miserable for women who needed to work. It also translated into unequal pay, in that women were excluded from skilled work wich paid more. Unwanted sexual aggression was of course also used against women slaves in the antebellum south. In the garment industry, it finally led to the entrance of women into the International Garment Workers Union, which was founded in New York in 1900. Yet, the union also rejected women whenever possible by claiming that men had to support families, while single women worked only until they found a husband.[49]

Men violently contested the right of women to earn more than minimum wages. Men would sabotage the work of women and smear oil on their coats or damage their property in other ways. Men generally refused to work if any woman ran a machine or was employed in skilled tasks. An "unwritten law" existed which would not allow women to do skilled work even if they were the bosses' relatives.Machine operating was reserved for men. The men who imposed these limitations on women refused to admit that these measures were discriminatory and pretended that segregation of work is "natural" or inborn.[50]

Nevertheless, by 1910 women began to use unions as their only means of countering sexual harassment, since there were no laws prohibiting this behavior. At the same time, some women in the garment industry needed to augment their starvation income by accepting gifts, meals and entertainment from men they met at work, particularly since marriage seemed to be the only means of escaping the sweatshop slavery.[51]

Sexual harassment consisted of demands for sexual favors in return for the pay the victim had already earned by her labor. Bosses would demand that female employees eat dinner and spend the night with them at a hotel.Male coworkers relieved their own misery in the sweat shops by inflicting daily indignities on women workers, such as speculating about the color of the women's underwear. Vulgar advances by male workers wee common features of everyday life for female sweat shop laborers, who often quit their jobs to escape the torture or bent their heads over their work.[52]

It was unionization which finally offered women a chance to resist sexual harassment. Between 1909 and 1913, Jewish and Italian immigrant workers as well as native born female workers joined the International Ladies Garment Workers Union in a series of strikes. The 1913 strike resulted in adding 50,000 new female members to the union. These women were adamant in their demands for better working conditions, which the male leaders of the union were unwilling to consider. Yet, as more and more women joined the union, they became the majority, so that by 1920 women comprised 75% of the membership[53]

Therefore, the 1909 strike and those thereafter led to the "uprising of the 20,000" who demanded an end to sexual harassment. They were aided by the Women's Trade Union League, a group of middle class and upper class women who had not worked in the garment industry but who used their ability to speak English and to influence the media to become bridges between the immigrant laborers and the American community. This was very important, because the media, the judges, and the politicians were almost unanimous in condemning strikes and using force to crush them. Some employers actually hired prostitutes to join the union picket lines in order to make it appear that female unionists were all "whores."[54]

While these strikes gradually improved working conditions and reduced, if not eliminated, sexual harassment, the income of women workers remained far below that of men. It is significant that in 1902 and again in 1915, after several strikes, New York State inspectors found that the wages paid women were only about half of the pay of men. Therefore, although women were the overwhelming majority of union members, skilled work and leadership positions in the union were almost all in the hands of the minority of men. In fact, through the 1920's, the ILGWU had only one female member on its executive board, Fannia Cohn.[55]

Thus, it wasn't until the law prohibited sexual harassment in 1964 that this conduct moved from the sweatshop into the labor unions. By then, the garment industry had begun to decline in this country, never to rise again.

The garment industry sweatshops were not confined to New York and other east coast cities. In the early 20th century, sweatshops also existed in the South and in the Midwest. Those in St. Paul, Minnesota, are one good example of these establishments. As in New York, the vast majority of sweatshop workers were young women who began their near slavery at age fourteen or younger. Unlike the east coast immigrants, these women were born in the United States. They came from the poorest families and were exploited accordingly by the textile manufacturers at that time. In 1926 there were 575 children in St. Paul, employed full time, sewing clothes, boots, and shoes. As a result, women employees founded a union, which became the "religion" of some of its members. This became true because those who left home to work in the factories were now forced to operate the power machines, which exhausted its users after a 14 hour working day. These laborers were paid $1.25 per day.[56]

It must be understood that the alternative to working in factories for women was work as a domestic servant. Such work was even more demeaning and did not allow the servants any time for themselves. Servants worked twenty-four hours a day, with some time off on Sundays. Other than limited employment opportunities, young women could only find support in marriage, which did not guarantee that they would not have to go to work at piecework in the home. Women who were divorced or widowed had no support other than factory labor or domestic servitude, as there was no Social Security law nor any other means of helping those in need. Then as now, there were some private social service agencies which collected money reputedly for the needy. However, then as now, the principal part of these collections became large salaries for "executive directors" and other sycophants with little left for those truly in need.[57]

The most prominent labor union in Minnesota in the early 20th century was the United Garment Workers. This union, even at the peak of its membership, enrolled only 10% of the women working in the sweatshops. This allowed employers to lock out union members who went on strike and to promote the "open shop." Together with the depression lasting from 1929 until the entrance of the United States into the 2nd World War, unions were largely powerless to help their members. It was the National Labor Relations Act of 1935 which finally gave unions the support needed to resist employers. Furthermore, it was the "Fair Labor Standards Act" of 1938 which finally gave American workers a minimum wage of $0.40 an hour and a 40 hour workweek.[58]

At least one reason wages were so low in the garment industry was that some manufacturers used prison labor, which cost them next to nothing. There were a number of states which had garment factories inside prisons. The prisoners were paid a fraction of union wages but the product was sold in the open market at the same price as union manufactured clothes. In 1929, Congress passed the Hawes-Cooper Act, which prohibited the sale of prison made products outside the state that manufactured it. That did not prevent the sale of prison made clothes within a state.[59]

In 1935, Congress passed the National Labor Relations Act, which guaranteed workers the right to organize and strike. Then, in 1938, the Fair Labor Standards Act provided a minimum wage, income for retirees, and set maximum hours of work at forty hours per week. All of these measures appeared to help the American worker until the 1990's, when globalization began leading to outsourcing on a massive scale and resulting in the loss of millions of American jobs to China and other countries. Therefore, the sewing machine became once more primarily domestic.

Nevertheless, there are still some American tailors left in New York. The youngest of these men are now sixty years old and many are more than seventy-five years old. These men are suit makers who learned the trade as apprentices in Italy before coming to the United States. Because technology has replaced human skill, only a few "master tailors" still work in the industry in the 21st century. There are, however, those who are employed at the highest level of suit making, which is still done by hand. Many of these skilled tailors come from China, Chile, or Eastern Europe, or are trained in the United States. As many as twelve languages were spoken in tailoring factories as late as 2011, andHickey Freeman in Rochester New York employs tailors from twenty countries.

According to the Bureau of Labor Statistics, there were 26,450 tailors in the United States in 2009. Two years earlier, in 2007, there were 31,550 people employed as tailors. The decline is considerable. The mean annual income of tailors in 2009 was $28,300 although in New York an annual wage of $34,330 reflected the high cost of living in that state.

V.

Government intervention was by no means the only cause for the decline in sweatshop conditions in the needle trades. The rise of unions and the willingness of the exploited workers to defend themselves was even more effective in fighting the horrors of exploitation in the garment industry. Beginning in 1909 and 1910, garment workers began to strike and were supported in these "uprisings" by middle class reformers who did not work in the "sweat" industries but who could not well tolerate the evident cruelties suffered by the garment workers. These middle class observers founded the National Consumers League in 1899 with a view of attaching a "White Label" to the garments sold in department stores. The "White Label" was to be attached to clothes made in factories that recognized unions and obeyed the labor laws. The label served to assure consumers that they were not buying clothes made in "sweatshops" and that child labor was not used.[60]

At the beginning of the 21st century, the apparel industry still represented 28 percent of all manufacturing jobs in New York City. This is in part true because at one time the garment industry was located almost entirely in New York. Furthermore, there are still many would be designers who believe that New York is the fashion capital of the world.According to the New York Times, it is still possible for an unknown designer to gain access to the fashion business in New York, although the chances of success are far less than was true in the middle of the 20th century. The Jewish women who worked in the sweatshops in the 1890's and early twentieth century are gone. Today, in 2011, workers in the garment industry come from Mexico and Ecuador, China and the Caribbean islands.[61]

Although the number of sweatshops gradually declined in the United States, such manufacturing practices still exist in the second decade of the 21st century. Therefore the New York Labor Department's Apparel Industry Task Force raided Suburban Textiles, Inc. on April 29, 2009, and confiscated all garments produced there. All equipment used to make these garments were also seized and kept until the manufacturer paid restitution to the underpaid women laboring there.The Labor Department found numerous violations of state law at the factory. These included failure to maintain payroll records, failure to pay overtime, assigning workweeks up to 80 hours, and violating state mandated rest periods. The outcome of that raid was the imposition of a fine of $500,000 owed to workers for wages and damages.[62]

In August of 2009, garment workers staged a protest against two clothing factories in Long Island City, New York. The cause of the protest was the failure of the owners to pay six Chinese workers who were owed hundreds of thousands of dollars in back wages and were then fired for demanding their money. The workers in the "sweatshop zone" on Long Island are mainly Chinese immigrants. Unlike their Jewish predecessors, these immigrants are better organized and have the backing of the New York State Labor Department. They also have the protection of the National Labor Relations Board, which did not exist until the passage by Congress of the National Industrial Recovery Act of 1933.[63]

The International Labor Rights Forum has developed a "Sweatshop Hall of Shame 2010." This list includes Abercrombie and Fitch, Hanes, Ikea, Pier 1, Wal-Mart, and others. According to that list, these companies employ laborers working long hours under dangerous conditions for poverty wages. The Labor Rights Forum accuses these and other companies of using suppliers who suppress workers' rights. It is further alleged that workers in Honduras who sought to form a union were dismissed and some were beaten.[64]

Evidently, sweatshops and the abuse of working people are still part of the garment manufacturing industry. However, unlike the years from 1890 to 1960, state investigations of these malpractices are effective and limit the abuses far better that was true in the early 20th century.

Sweatshops became world wide monstrosities at the end of the 20th century. This is particularly true in China as well as other Asian countries. There, workers come from rural villages to cities looking for work and end up slaves to the garment industry. The workweek in these sweatshops may be more than 70 hours, even if the law limits the work week to 49 hours. Large corporations make clothes for Wal-Mart in these areas. They evade overtime laws and restrict the workers' use of the companies' medical facilities. These companies seek to reject unions and ignore the health and welfare of their workers.

China has developed thousands of factories where migrants from rural areas slave at producing goods for western companies, particularly Wal-Mart and other American enterprises. These factories are ruthless in their exploitation of workers. Although such factories pay only $22 a month, they charge $15 for food per month. These factories also give workers "residence permits" which allow them to live in company towns. Workers must give management their personal identification card, without which a Chinese subject cannot go anywhere. This makes the workers virtual captives of management. These factories also employ guards who beat workers for talking back to management. Factories in China do not pay overtime. There are labor laws in China which are seldom observed, so that the sewing machine continues to enslave millions.[65]

VI.

Although the sewing machine enslaved thousands of women over more than a century, there are also those who view the sewing machine as a means of enjoyment and income. The New York Times published a story about "the Williamsburg seamstress," who is a traveling tailor living in Brooklyn, New York.Using a website, Ms Nayantara Banarjee gains clients whom she visits at any time convenient to both. She visits some clients late at night and others in the morning, as she hems curtains and skirts, shortens sleeves, and makes men's suits.

Bon in India and raised in the U.S., Banarjee learned sewing from her mother beginning when she was six years old. In business for herself, she is no sweatshop slave, but makes a decent living and enjoys her work.[66]

There are those to who view sewing as an enjoyable hobby and who use the sewing machine creatively. These are almost all women who have learned to work with the tools needed to make clothes or other items such as bed covers, curtains, and quilts. These tools include dressmaker's shears, stick pins, sewing chalk, sewing needles, seam rippers, and hard rulers. There are now machines that can do embroidery and make buttonholes, although many sewing enthusiasts make these things by hand.[67]

Because sewing has many followers who view it as a means of relaxation and enjoyment, there are numerous books in any library dealing with various aspects of the sewing enterprise. There is The Complete Photo Guide to Sewing by Editors of Creative Publishing; Sew and Stow by Betty Oppenheimer; The Encyclopedia of Sewing Machine Techniques, and Couture Sewing Machine Techniques by Claire B. Shaeffer. "Amazon" sells over eight thousand books dealing with sewing, so that we have good reason to speculate that the sewing machine will constantly expand.[68]

There are also schools which teach sewing to adults. Some of these schools are "on line" schools such as "Sew Fast and Sew Easy," which teach classes in Beginning Sewing, Fashion Sewing, Accessory Sewing, and making skirts, pants, and tops. Crotchet knitting is also taught, as are garment industry techniques. The first course taught by "Sew Fast" takes ten weeks and costs $348 tuition. The school promises that in five sessions enrollees will learn how to use commercial sewing patterns, how to thread and use all sewing machines, sew up and envelope a pillow, how to lay out and cut patterns, how to use "7th Avenue" sewing skills, how get your own custom fit in your skirt, how to identify different types of fabric, pattern and sewing terms, special pressing techniques, and how to make and sew hems.[69]

Because New York has been the center of the needle trades and the garment industry, the city refers to "7th Avenue" as "Fashion Avenue" on street signs.There is in New York a "fashion district" where fashion "banners" are displayed and visitors stroll along the "Fashion Walk of Fame." This includes a statue of a man bending over an old sewing machine alongside the Fashion Center Information Booth, which displays a large button on its roof. The entire display was opened in 2000 with much publicity.[70]

VII.

Because sewing is both a hobby and has commercial purposes, there are numerous books and magazines in print concerning sewing. Among these is Altered Couture Magazine, which includes numerous photographs of jackets, skirts, sweaters, jeans, and trousers, altered and enhanced. There is also Handwoven Magazine and Sew Beautiful Magazine. This magazine teaches both machine and hand sewing, as does Sew News Magazine and Threads Magazine. There are a good number of other sewing magazines and a number of books such as Super Simple Sewing and Sewing School aimed at children old enough to operate a sewing machine.There is also an Encyclopedia of Sewing Machine Techniques which includes different methods of drawn string techniques, decorating bobbin work, quilting, and ribbon techniques.

Barbara Burman has edited a book called The Culture of Sewing. This is sociology of sewing and includes chapters on Dynamics of Class and Gender, House Dressmaking in Edwardian England, History of Making Clothes at Home, Making the Modern Woman, and Home Sewing in the 20th Century.

Burman and other contributors to her book have shown that home sewing has a very long history, as it was an almost universal experience in the lives of women in the United States and elsewhere. Prior to the advent of the second half of the 20th century, it was routine to teach girls to sew because it was a practical necessity. As ready made clothes became the rule, home sewing became a hobby for many women. Sewing is of course an aspect of material culture, as can be seen by reviewing the technological developments affecting the sewing machine. Sewing is also an aspect of behavioral culture, as it is both a means of earning a livelihood and a hobby for a considerable number of women in the United States and Europe. Thirdly, sewing is an aspect of ideological culture, as it relates to the belief that sewing at home is for women and therefore connects to the entire gender identity of women, as imposed by the age old patriarchy only being challenged in the 21st century.

Summary

Before the invention of the sewing machine, women were enslaved by constant and endless sewing. The sewing machine was viewed as a "great liberator" of women. When the industrial sewing machine led to the establishment of "sweatshops," immigrant women became the victims of gross exploitation by brutal garment factory owners who even used child labor. Eventually working men and women organized labor unions that used strikes to gain better treatment, which was finally accorded them because of laws passed during the Franklin Roosevelt administration. Since the 1960's, the garment industry has rapidly declined in the United States because of "outsourcing" and because China and other countries once more use "sweatshops" to produce the clothes worn by Americans and others.

Notes

1. Sophia Skoric and George Vid Tomashevich, *Serbs in Ontario: A Socio-cultural Description,* (Toronto: Serbian Heritage Academy, 1974), p.69.

2. Sarah Hale, "The Seamstress," *Godey's Lady's Book,* vol 74, (1867).

3. Thomas Hood, "Song of the Shirt," In: R.B. Inglis, *Adventures in English Literature,* (Toronto: W.J. Gage, 1952), pp 436-437.

4. Amy Boyle Osaki, "Truly Feminine Employment," *Winterthur Portfolio,* vol. 23, no. 4, (Winter, 1988), p. 225-241.

5. Grace Rogers Cooper, *The Invention of the Sewing Machine,* (Washington, DC: The Smithsonian Institute, 1968), Chapter 1.

6. Elizabeth M. Bacon, "Marketing Sewing Machines in the Post Civil-War Years," *Bulletin of the Business Historical Society,* vol. 20, no. 3, (June 1946), pp. 90-91.

7. Grace Rogers Cooper, *The Invention of the Sewing Machine,* (Washington, DC: The Smithsonian Institution Press, 1968), p.15.

8. Robert McIntosh, "Sweated Labor: Female Needleworkers in Industrialized Canada," *Labour,* (Fall, 1993), pp. 105-138.

9. Wally Seccombe, "Patriarchy Stabilized: The Construction of the Male Breadwinner Wage Norms in the 19th Century," *Social History,* vol. 11, no. 1 (January 1986), pp. 66-67.

10. Jenny Morris, "Women Workers and the Sweated Trades: The Origin of Minimum Wage Legislation,"(Kansas City: Brookfield. Publishers, 1986), pp. 192-194.

11. Mary Ann Poutanen, *For the Benefit of the Master, 1820-1842,* (Montreal: McGill University, 1985).

12. Gerald Tulchinsky, "Said to be a Very Honest Jew," *Urban History Review,* vol. 18, no. 3, (February 1990), p. 206.

13. James Schmiechen, *Seated Industries and Sweated Labor,* (Urbana, IL: University of Illinois Press, 1984), p. 2.

14. Michael S. Cross, *The Workingman in the 19th Century,* (Toronto: Oxford University Press, 1974), p.153.

15. Roger D. Waldinger, *Through the Eyes of the Needle: Immigrants and Enterprise in New York Garment Trades,* (New York: New York University Press, 1986), pp. 54-55.

16. Mercedes Steedman, "Skill and Gender in the Canadian Clothing Industry, 1890-1940" in: Craig Herron and Robert Storey, eds. *On the Job: Confronting the Labor Process in Canada,* (Montreal and Kingston: McGill and Queen's University Press, 1986), p. 158.

17. *Annual Report of the Bureau of Industry,* (Toronto: 1889), p. 10.

18. Wyllistine Goodsell, *A History of Marriage and the Family,* (New York: The Macmillan Co., 1934), p. 497.

19. Mackenzie King, "Foremen demanding bribes," *The Daily Mail,* (Toronto: October 9th,1897), p. 10.

20. Carla Lipsig, "Organizing Women in the Clothing Trades:" *Studies in Political Economy,* vol. 22, (1987), pp. 41-71.

21. Christine Stansell, *City of Women: Sex and Class in New York,* 1789-1860. (Urbana, IL: The University of Illinois Press, 1987), p. 117.

22. Susan Wortman, "The Unhealthy Business of Making Clothes," *Healthsharing,* vol. 1, no. 1, (November 1979), p. 12.

23. James Schmiechen, *Seated Industries,* p. 61.

24. No author, "History of American Agriculture, 1776-1990," *Farmers and the Land,* (New York: The New York Times Company), Time line.

25. Lynn Anderson Rieff, *Rousing the People of the Land,* (Auburn, AL: Auburn University, 1995), p. 50.

26. U.S. Department of Labor, "Usual Weekly Earnings of Wage and Salary Workers, Fourth Quarter, 2010," (Washington, DC: Bureau of Labor Statistics, 2011), Table 9.

27. Christine Stansell, *City of Women: Sex and Class in New York,* pp. 113-114.

28. Burton Hall, "I.L.G.W.U.: Its Enemies and Its Friends," *New Politics,* (Fall, 1976), p. 46-50.

29. Jenny Morris, *Women Workers and the Sweated Trades, The Origins of Minimum Wage Legislation,* (Aldershot. England: Hants Publishing), Ch. IV.

30. U.S. House of Representatives, 52nd Congress, Second Session, Committee on Manufactures, "The Sweating System," vol. 1, no.1, (Washington, DC: The United States Government Printing Office, 1893), pp. iv-viii.

31. No author, "The Story of the Sewing Machine," *The New York Times,* (January 7, 1860), p. 2.

32. No author, "The Old Issues Again," *Sewing Machine Times,* (December 25, 1902), p. 8.

33. Susan Strasser. *Never Done,* (New York: Henry Holt & Co., 2000), p. 134.

34. Robert S. Lynd and Helen Merrill Lynd, *Middletown: A Study in Contemporary American Culture,* (New York: Harcourt, Brace and Co., 1929), p. 3.

35. United States Department of Labor, Bureau of Labor Statistics, *Minimum Quantity Budget Necessary to Maintain a Worker's Family of Five at a Level of Health and Decency,* (Washington, DC: Government Printing Office, 1920), p. 20.

36. Claudia Kidwell and Margaret C. Christman, *Suiting Everyone: The Democratization of Clothing in America,* (Washington DC, Smithsonian Institution Press, 1974), p.1.

37. Ethel Lloyd Patterson, "The Cost of Being Well Dressed," *Ladies Home Journal,* Vol. 40, no. 3. (April 1923), p. 88.

38. Editorial, "The Fire That Changed Everything," *The New York Times,* February 23, 2011, p. A22.

39. Leon Stein, *Out of the Sweatshop: The Struggle for Industrial Democracy,* (New York: Quadrangle/New Times Book Co., 1977), pp. 194-195.

40. Lara Lemlich, "Life in the Shop," *New York Evening Journal,* (November 28, 1909).

41. Dermott Quinn, *The Irish in New Jersey,* (New Brunswick, NJ: 2004), pp. 61-149.

42. Jay Michael Hollender, "Prelude to a Strike," *Proceedings of the New Jersey Historical Society,* vol. 79, (July 1961), pp. 161-168.

43. Gerald Markowitz and David Rosner, *Slaves of the Depression: Workers' Letters About Life on the Job,* (Ithaca: Cornell University Press, 1987), pp. 76-80.

44. Monty Finiston, ed., *Oxford Illustrated Encyclopedia of Invention and Technology,* (Oxford, England: Oxford University Press, 1992).

45. Ibid., p. 46.

46. Marc Weingarten, "To Stitch, Take Needle, Thread and a Computer; New Machines Revolutionize Sewing," *The New York Times,* (August 27, 2004), p. 14.

47. Jessica Figer, "Twilight of the Sewing Machine Repairman," *Metropolis,* (February 11, 2011), p. 1.

48. Mary Bularzik, "Sexual Harassment at the Workplace," In: James Green, Editor, *Worker's Struggles, Past and Present, I* (Philadelphia: Temple University Press, 1983), pp. 117-135.

49. Katherine Franke, "What's Wrong with Sexual Harassment?" *Stanford Law Review,* vol. 49, (1997), pp. 771-772.

50. Ruth Milkman, "Redefining Women's Work: The Sexual Division of Labor in the Auto Industry During World War II," *Feminist Studes,* v. 8, (summer 1982), pp. 337-372.

51. Alice Kessler-Harris, "Organizing the Unorganizable: Three Jewish Women and Their Union," *Labor History,* vol. 17, (Winter 1976), p. 211.

52. Joan Wallach Scott, *Gender and the Politics of History,* (New York: Columbia University Press, 1988), pp. 93-112.

53. Alice Kessler-Harris, "Where are the Organized Women Workers?" *Feminist Studies,* vol 3, (Fall 1975), pp. 92-110.

54. Editorial, "Contempt of Court or Contempt for a Vicious System? Jail Sentences on Frail Women," *The Ladies Garment Worker,* January 1912), p. 1.

55. Annelise Orleck, *Common Sense and a Little Fire: Women's Working Class Politics in the United States,* (Chapel Hill: The University of North Carolina Press, 1995), pp. 72-74.

56. Minnesota Bureau of Labor Statistics, *Eighth Biannual Report, 1901-1902,* p. 462.

57. Louis Levine, *The Women's Garment Workers: A History of the International Ladies Garment Workers Union,* (New York: B.W. Huebsch, 1924), p. 342.

58. Public Law 75-718, ch. 676, 53 Stat 1060, June 25, 1938, 29 USC ch. 8.

59. No author, "Garment Manufacturers meeting in New York City," *Minnesota Union Advocate,* (January 21, 1925), p. 6.

60. Robert J.S. Ross, *Slaves to Fashion: Poverty and Abuse in the Sweat Shops,* (Ann Arbor: The University of Michigan Press, 2004), p. 86.

61. Guy Trebay, "Needle and Thread Still Have a Home," (*The New York Times,* April 28, 2010), p. E1.

62. Leo Rosales, "State Raids New York City Sweatshops," Albany, NY, *New York State Department of Labor,* (http://www.labor.state.ny.us/pressrelease/2009/April29-2009/htm.

63. Adrianne Pasqaurelli, "Garment Workers Protest Sweat Shops in Long Island City," *Workforce Management,* (August 12, 2009), p. 1.

64. Trina Tocco, "Sweatshop Hall of Shame 2010," *International Labor Rights Forum,* (no date), pp. 1-2.

65. Dexter Roberts amd Aaron Bernstein, "Inside a Chinese Sweatshop: A Life of Fines and Beatings," *Bloomberg Business Week,* (October 2, 2000), p. 1.

66. Gillian Reagan, "Stitch by Stitch and Block by Block," *The New York Times,* (February 17, 2011): Style Desk, p. 1.

67. Janelle Dailey, "Sewing is Relaxing and Productive," *Countryside and Small Stock,* (June 2005), p. 96.

68. http://www. amazon.com.

69. http://www.sewfastseweasy.com.

70. Debbie Colgrave, "A Walk You Don't Want to Miss," http://sewing.about.com/library/weekly/aa101301a.htm (2001).

Chapter Five

The American Gun Culture

I.

"A well regulated Militia, being necessary to the security of a free State, the right of the people to bear Arms, shall not be infringed."[1]

This amendment has led to endless controversy among those who believe that the right to bear arms was only valid as long as a citizens' militia was operative in the United States and those who ignore the first clause of this amendment.

Because this amendment causes so much controversy, the Supreme Court of the United States has been called upon to decide the meaning of this amendment. The court has ruled on this issue several times. One of these rulings, United States v Miller in 1939, was inconclusive and caused a great deal of confusion concerning the meaning of the 2nd amendment. Later, in District of Columbia v Heller in 2008, the court ruled that "the hand gun ban ... violates the second amendment" and told the District that it "must permit Heller to register his handgun and must issue him a license to carry it in the home."[2]

In 2010 in McDonald v. Chicago, the Supreme Court once more ruled in favor of a citizen to "keep and bear arms."The court held that this right is protected by the second amendment and the fourteenth amendment of the constitution.

Richard Uviller and William Merkel have argued that the second amendment applied only to a militia armed by the federal government in the early days of the republic and those such militias no longer exist. They conclude that the right to bear arms does not arise from the second amendment to the constitution.[3]

The ruling of the court is of course the law, while the views of two professors are a private matter, not enforceable. It is evident that those who seek to invalidate the second amendment would also seek to invalidate the entire Bill of Rights if they had succeeded in persuading the Supreme Court to hold with them. Moreover, the possession of firearms by the American citizenry is the best protection Americans have to insure that government never becomes a dictatorship, as experienced in innumerable societies over the whole of mankind's history.

It has been estimated that about four in ten Americans have a gun in their home. This means that around 120 million American adults own about 290 million firearms. Forty percent, or 116 million, of these guns are handguns and cannot be used for hunting. These handguns consist of revolvers, semi-automatic pistols, and a few "other" handguns.

Gun ownership is unevenly distributed in the United States. In the Midwest, 34% of adults own a firearm, and in the South, 36% of adults own a firearm. Nearly half of American men, i.e. 47%, own a firearm, as compared to 13% of women.[4]

It is instructive to compare the rate of homicides by weapons used by the perpetrator. This reveals at once that two thirds of killings in America are carried out by a firearm. According to the Federal Bureau of Investigation, there were 13,636 murders known to the police in 2009. Forty-seven percent of these murders were carried out by handgun, nearly three percent by using a rifle, three percent of killings involved a shotgun, and in another fourteen percent of killings an unidentified firearm was used. Clearly, firearms are the preferred instrument for homicide in the United States.[5]

There can be no doubt that the rate of killings and other crimes has declined precipitously in the USA over the past thirty years. In 1980, when the U.S. population was slightly more than 225 million, the rate of homicides was 10.2 per hundred thousand inhabitants. In 1990, the population had increased to 249 million, with a homicide rate of 9.4 per hundred thousand, and in 2010, with a population exceeding 308 million, the homicide rate has declined to 5.0 per hundred thousand.[6]

These statistics demonstrate that the possession of firearms by the American people cannot be the reason for homicide. No doubt, homicide is facilitated by firearms, but not caused by firearms. The decline in homicide and other violent crimes in the United States must be found elsewhere.

If we consider that over eighty percent of murders are committed by men between age 15 and 45, of which fifty-nine percent are attributed to men between 15 and thirty years of age, then it is evident that homicide is not caused by the presence of guns. This is further supported by the decline in the American birthrate, which was 16 per 1,000 population in 1988 and then fell to 14 in 2004, a period of sixteen years. In 2010, the birthrate had declined to 13 per thousand population. Therefore it is evident that the rate of homicide declined in part because the number of men in the age group most prone to killing others has declined. In short, murder is less because there are fewer murderers. Nevertheless, the gun is the most important contributing device to homicide in the country.[7]

School shootings undoubtedly have received the most publicity of all homicidal acts in the United States. The media, ever ready to exaggerate for the sake of profit, leave the impression that youthful gangs have sophisticated weapons, including military assault rifles.[8]

There are young people with such weapons. These are generally youths involved in the drug wars or are members of gangs. Most armed young people do not have such firepower but instead own a handgun. Those under age 21 cannot buy such firearms legally because the Gun Control Act of 1968 prohibits gun sales to those under twenty-one.[9]

So-called "juveniles" carry guns because they believe they need a gun for self defense or to commit aggression or to enhance their status among their peers. Criminologists estimate that sixty-three percent of firearms acquired by "juveniles" are handguns such as semi-automatic pistols and revolvers, and that juveniles are much more likely to possess non-powder firearms, while adults usually possess high powered rifles and shotguns.[10]

These guns include so-called "Saturday night specials," which are defined as .32 caliber or less and having a barrel length of 3.1 inches or less. The guns are inexpensive, as they retail for $100 to $150 and are even less on the street. The cheap price makes these guns popular with youths, who usually don't have the money to buy more expensive arms.

The National Firearms Act and the Gun Control Act have made several kinds of guns illegal. Illegal are rifles and shotguns that have been altered by sawing off the barrel so as to make it easier to conceal them. Likewise, removing or altering the serial numbers on guns is illegal. Every state in the Union has gun laws, so that there can be differences between states as to what it or is not allowed.[11]

The Omnibus Crime Bill which Congress passed in 1994 banned assault weapons importation into the United States, including some semi-automatic firearms. These are rarely used by young people, who prefer to use BB guns or pellet guns.These guns have relatively little capacity for injury, although in a few cases they have been used to kill. Even if that is unusual, it is most likely that pellet guns cause serious wounds and also cause a great deal of pain.[12]

II.

Guns are of course produced by gun manufacturers. There are about 110 gun producers in the United States. These gun manufacturers employ a powerful lobby, influencing Congress and other politicians to limit the law pertaining to guns. The manufacturers are not all alike. In fact, only ten manufacturers of guns dominate the market.

The gun market is divided into a primary market consisting of federally licensed retailers including pawn brokers and a secondary market consisting of private persons selling guns on occasion and unlicensed vendors at gun shows.[13]

It is common for licensed retailers to sell guns in bulk by way of surrogates to private unlicensed parties, who then resell them to anyone, including youthful street gangs. This means that retailers may well be the source of crime guns, either directly or indirectly. Youth under age 18 are known to go out of state to buy guns in quantity. In addition there are so called "straw" purchasers. These are people who buy guns for those prohibited from buying guns, including those convicted of felonies, mentally ill people, and underage individuals.

Because the federal law concerning gun dealing is vague, there are innumerable unlicensed gun dealers who claim they are pursuing a hobby and therefore claim they are not "engaged in the business" of gun dealing, and accordingly do not need a license. At gun shows, vendors do not need to ask purchasers for identification, although they may ask the age of the buyer without proof.[14]

Guns are not only bought but also stolen. It has been estimated that about 500,000 guns are stolen each year, although the theft of guns rarely involves the semi-automatic pistols criminals generally use. This is true because pistols hold more ammunition than revolvers, so that operating them is too expensive for most youths. Revolvers hold only six bullets, while the "double stack" magazines of pistols hold 20 rounds or more. Gun caliber has also increased in the decade ending in 2010, so that guns used in a civilian setting are more lethal than ever before.[15]

There are those who claim that the gun industry deliberately increased the fire power of new guns so as to sell them to those already saturated with less powerful guns.

In 1994, Congress passed the "Brady Handgun Violence Prevention Act," which requires a five day waiting period prior to a handgun purchase. That law was named after James Brady, who was severely wounded when John Hinckley shot President Ronald Reagan in 1981. Brady was Reagan's press secretary. Both Reagan and Brady survived, although Brady has been confined to a wheelchair for over thirty years.

The Brady Act was amended in 1998 and was thereafter called The National Instant Criminal Background Check System. More than ninety percent of these required background checks are now completed within two hours of application, and 72% are completed in 30 seconds. The law allows retailers to release guns to applicants within three days, even if the background checks are not completed. Not all states operate under the Brady system, so that different states have a variety of gun control laws.[16]

A number of suggestions have been made designed to curb the use of guns in the U.S.A., although none have proved effective. Among these is the regulation of licensed retailers, screening prospective buyers, limiting gun sales, regulating the secondary gun market, registering guns and licensing owners and banning weapons of choice.

III.

The possession of guns is a political issue in the United States. Because the second amendment provides Americans with the right to own firearms even as opponents of that amendment have for years attempted to curtail or even eliminate that right, gun owners have organized the National Rifle Association.

That Association is devoted to keeping the right to own firearms legal in the United States.

The National Rifle Association was founded in 1871 when two former Union army officers, Colonel William Conant Church and General George Wingate, formed the NRA in order to foster marksmanship. The NRA increased its membership considerably between the two world wars, as the U.S. government gave the NRA leftover rifles from the wars at cost. The NRA then developed a Legislative Affairs Division to deal with gun legislation nationwide, but with particular emphasis on Congress. The NRA also started a publication, "The American Rifleman."

In 1939, the NRA collected 7,000 guns to help Great Britain defend itself against Germany. In 1941, when the U.S.A. entered the Second World War, the NRA guarded factories and other installations involved in the war effort. Together with its concern with hunting and police marksmanship, the NRA reached a membership of 300,000 in the 1950s.

In the 1970's, the NRA acquired 37,000 acres of land in Arizona and launched a new publication, "The American Hunter." By vigorously opposing gun control in several states, the NRA raised its membership to one million in 1975 and 2.6 million in 1981. By 1986, the NRA had three million members. In 1997, the NRA started publishing "The American Guardian."In 2000, the NRA had 3.7 million members. In 2011, the NRA had 4 million members and therefore was able to powerfully influence legislators seeking re-election. The actor Charlton Heston (John Carter 1923-2008), became president of the NRA in 1998 and remained in that office until 2003.[17]

The National Rifle Association also runs a museum at the NRA's headquarters in Fairfax, VA. This $2 million exhibition includes so-called "Tommy guns," invented by General John T. Thompson in 1919. This is a submachine gun which was used extensively by organized criminals and the police during the prohibition era. It was also used by American forces during World War II in the form of the "M1."[18]

A much smaller organization concerned with firearms regulations is the JPFO or "Jews for the Preservation of Firearms Ownership." This group, headquartered in Houston, Texas, remembers the Holocaust, in which six million European Jews died at the hands of German and other European persecutors. The JPFO recalls that in 1919, fourteen years before the Nazi takeover of the German government, the German republic enacted a law requiring all Germans to hand over all firearms to the government. This meant that the Nazi government had no trouble dictating to all Germans, as no one had the wherewithal to resist. This is not to say that the Holocaust would not have happened if guns had been available. It does mean, however, that some resistance might have been possible. Unfortunately, "the only thing we learn from history is that we learn nothing from history," so that the Nazi episode is now forgotten, except for a few surviving victims.[19]

All this enthusiasm for gun ownership has led many observers to the conclusion that there is a "gun culture" in the United States. This means that many Americans and others believe that the history of the country has produced this gun culture. It is often written and widely assumed that this gun culture began early in America and was caused by the necessity of those who settled the west to defend themselves against Native Americans and each other.

In an exhaustive study of this belief, Michael A. Bellesiles has rejected this thesis. Although Bellesiles does not deny the present existence of the gun culture, he has demonstrated that guns were seldom owned, rarely used, and usually not very effective before the Civil War (1861-1865), and that guns became popular as the consequence of American industrialization and not because of the settling of the western frontier. It was hunting and militarism which created the gun culture in this country, particularly after the Mexican War (April 1846 - February 1848) led to a large scale production of guns and the enthusiasm of the public for their use.[20]

The Mexican War resulted in the invasion and annexation of Texas, California, New Mexico, Utah, Arizona, and Nevada. These additions to the United States were largely the result of the occupation of Mexico City by an American force under General Zachary Taylor (1794-1850), who became the 12th President of the United States. Taylor proved that guns were indeed effective in gaining one's ends, thereby encouraging the gradual development of the gun culture in this country.[21]

Since the end of the Mexican War, the United States has been continuously involved in wars against numerous native American tribes, foreign wars such as the Second China War from 1856-1859, the war against Paraguay, also in 1859, the Civil War, from 1861 to 1865, the Sioux Uprising of 1865 followed by additional campaigns against native Americans, lasting until 1891. There followed the Spanish American War in 1898, campaigns against Mexico, Nicaragua, China, and the Phillipines, the First World War in 1917-1918, intervention in the Russian civil war in 1919-1920, the Second World War, 1941-1945, the Korean War 1950-1953, the Viet Nam War 1960-1975, the invasion of Panama in 1989-1990, and the first Iraq war 1990-1991. The United States became involved in the wars in Somalia, Bosnia, and Kosovo, 1992-1995, and the second Iraq war including Afghanistan beginning in 2001 and continuing for ten years or more. In short, the United States has been at war during almost all the years since 1848, so that innumerable men, and some women, serving in the armed forces during all these years became familiar with firearms and began to keep them at home. This is of course particularly true of law enforcement personnel.[22]

IV.

War became a source of considerable income for gun manufacturers in the United States because the U.S. government ordered only American made firearms. This was not true of private hunters, who persisted in ordering their guns from Europe. It was mainly through the efforts of Samuel Colt (1814-

1862) that handguns became popular in the United States, after Colt invented the "revolver" or "revolving gun" in 1832, when he was eighteen years old.[23]

Colt built a large factory in Hartford, Connecticut, in 1854. Using mass production techniques and producing interchangeable parts, the Colt revolver was sold all over the world, so that Colt owned the largest private armory anywhere. During the Civil War, (1861-1865), Colt sold guns to the U.S. government, making the revolver even more popular. Colt had also set up a firearms factory in New Jersey, where he produced rifles and shotguns. Colt's guns were widely used by the pioneers who moved west after the Civil War.Colt died at the age of 47, but his legend and success have continued, as the Colt Firearms Co. celebrated its 175th anniversary in 2011.

During the 2nd World War, the Colt factories employed 15,000 people, only to lose money after the war. Subsequently the Colt Co. merged with a holding company and thereafter suffered a number of business setbacks, only to return to success after the U.S. government ordered thousands of carbines and M16 rifles from the company. The company also became successful in the sporting goods business, as well as making handguns for police departments around the country. These sales and the availability at gun shows of handguns and other firearms are a part of the gun culture in America in the early 21st century.[24]

There are 110 firearms manufacturers in the United States, who produce about 1.4 million small arms each year. Only about 37,000 of these small arms are sold to the military each year, so that the overwhelming number of handguns of all kinds are in private possession, distributed within the American population.[25]

An advertisement on the internet displays the cost of guns in the U.S. in 2011. Accordingly, antique guns which are collector's items are quite expensive. A Remington handgun dating back to 1890 sells for $7,500, while the majority of Remington pre-1899 handguns cost about $1,500. Other more modern handguns cost less. A Beretta sells for $540 to $750, which is also true of Glock handguns, although a Colt "Century" sells for $1400.The well known gun manufacturer Smith and Wesson has a popular pistol selling at $1,050.

Also for sale on the internet are rifles. Winchester rifles cost between $688 and $2,320, although some Ables rifles may be had for $400. Ables also sells "Sabers" defense rifles, which are semi-automatic and cost between $1,100 and $1,800. Those citizens who need a machine gun can buy one or more on the internet for a bit more than $4,000 or spend as much as $18,500 to be more effective.[26]

Guns are also sold at gun shows, which operate in every state of the union all year. Included are antique guns, which are collected by members of numerous associations with particular interests in guns of various origins. "Antique and Collectible Firearms and Militaria Headquarters" appears to be the largest of these antique gun collection groups. There is also the Cowboy Mounted Shooting Association. The members shoot black powder blank cartridges at balloons located between barrels while riding a horse. The guns used are single action revolvers like those used in the old west. Then there is the

National Muzzle Loading Rifle Association, whose members engage in historical reenactments. The Civil War Guns, The Colt Collectors Association, The Smith and Wesson Collectors Association, and The Winchester Collectors are the names of some more of these enthusiasts.[27]

Some gun collectors' associations are based on geography rather than history. There are gun collectors' associations in every state of the union. The dues of these associations are in the vicinity of $50 a year and anyone is welcome to join. The associations publish magazines like "The Texas Gun Collector."

Gun collectors are constantly involved in disputes with those who believe that crime, and particularly murder, is caused by gun owners. Collectors keep repeating that "guns don't kill people, people kill people."Anti-gun advocates believe that crime is caused by the possession of guns, particularly handguns. Gun owners find it ridiculous to claim that guns, which are inanimate objects, are viewed by the anti-gun crowd as having the miraculous power to shoot and kill as if the gun has a mind of its own.[28]

Opponents of gun ownership hold the view that some of the most spectacular mass shootings in this country could never have taken place were it not for the easy availability of guns. This can hardly be denied, unless it could be shown that those who commit such heinous acts would have had another means of achieving the mass murders they perpetrated.

Among those who committed extreme violence by means of one or more guns was James Oliver Huberty. On July 18, 1984, Huberty walked into a McDonald's restaurant in San Ysidro, California, and killed 21 people at random. In addition, he wounded 16 people. He used three guns with which he fired at restaurant guests and employees indiscriminately. Huberty aimed "a relentless stream of gunfire" at the occupants of the restaurant until bloody corpses lay everywhere. Some were on the counter, others on the floor or across the tables. The dead included several children, some of whom were outside the restaurant and whom he shot from the inside. The rampage continued in the town on the Mexican border until a police sharpshooter killed Huberty, who was then 41 years old.[29]

There are of course many Americans who are as well armed as was Huberty in 1984. Yet hardly any of these gun collectors have ever shot anyone. It is therefore evident that an argument can be made for the view that the shooter, not the gun, is responsible for mass murder.

Huberty's parents divorced when he was still in grade school. After his father remarried, he seldom saw him. As an adult, he kept to himself, but did marry and raise two children. He was angry all the time. He argued with neighbors and threatened them with attack dogs and his guns. Huberty was a gun lover and was a college graduate. He first worked as an undertaker but was soon fired because of his failure to "relate to people," according to his erstwhile boss, Don Williams. He then went to work as a welder, but lost that job when his employer, Babcock and Wilcox, closed down for lack of orders. He then decided to move to Mexico, and ended up in San Ysidro, where he found a job as a night watchman. He soon lost that job, and having told his wife, "I am going to hunt humans," murdered 21 people in anticipation of committing "suicide by cop." [30]

In the American competitive world, those who view themselves as "losers" can become so deprived that they are willing to die in a spectacular manner after leaving out their anger at the world by killing others. This means that not the gun, but the perpetrator, is responsible for firearm crimes.Evidently, Huberty was not the only mass killer whose feeling of deprivation led to such horrible deeds. There were also Leslie Lowenfield, James Schnick, Carl Isaacs, Charles Whitman, and more recently, Eric Harris and Dylan Klebold.

Furthermore, there have been other mass murderers who did not use any firearm. Most remembered among them was Timothy McVeigh. McVeigh used fertilizer to make a bomb that killed 168 people and injured 680 people at the federal office building in Oklahoma City on April 19, 1995. McVeigh wrote in his autobiography that he viewed himself as a "loser" and had a strong wish to die. Evidently, American culture produces a good number of people who have such feelings, although they are seldom mass murderers. [31]

It is not unreasonable to speculate that some gun collectors are motivated by the need to feel powerful in a world in which more and more Americans feel powerless in view of the ever encroaching government bureaucracy, the endless rules and regulations impinging on American citizens, and the progressive dilution of those rights and freedoms listed in the "Bill of Rights" of the Constitution. Therefore, not guns but relative deprivation and economic failure are far more responsible for violence in this country than the mere possession of guns.

V.

Hunting is a sport. At least, its advocates say so. Nearly 12.5 million people hunt animals in the USA. These hunters and anglers spend $34 million in any hunting and fishing season to the benefit of such hunting equipment providers as the Bass Pro Shops, who conduct a 17 day "Classic" in August of each year. That "Classic" consists of a series of seminars at which hunters are told by "experts" what strategies to use in the field to hunt all kinds of animals and what kind of gun to use. Hunting and fishing supports about 1.6 million jobs in this country and a billion dollars on tags, licenses, and permits. In addition, federal, state, and local taxes generate another $25 billion each year. [32]

Guns used for hunting are either rifles or shotguns. Rifles have a grooved bore, shoot a single projectile, are long range, and have front and rear sights. Shotguns have a smooth bore, shoot multiple projectiles, are short range, and have a front sight only.Rifles can cost $173 or as much as $1,455 and average around $650. A shotgun can cost as little as $325 or as much as $12,000 with some a good deal cheaper. There is also a deluxe shotgun advertised as $49,995 because it has gold inlays and an engraving by a well known artist. [33]

These weapons are used for "wildlife management," as hunting enthusiasts like to call their sport. The reason for "wildlife management" is that the deer population in urban areas needs to be reduced. Large deer populations in urban areas are associated with Lyme disease, deer-vehicle accidents, and damage to plants. Although firearms hunting has been used successfully to reduce deer populations, there are a good number of gun use opponents who believe that firearms used in hunting are a safety concern. Some states have therefore decreed shooting proficiency tests or have imposed interviews with prospective shooters applying for a license, while others have restricted hunting to certain hours.[34]

The fact is that hunting is extremely important to wildlife conservation in the United States. Furthermore, hunters contribute millions of dollars by buying hunting licenses and paying taxes on hunting equipment. In the rural past of this country, hunting was part of every American's life and was taken for granted as a yearly event. Today, in 2012, the population of rural America is only 16%, down from 72% one hundred years ago. Therefore, hunting is declining in the United States, so that the International Association of Fish and Wildlife Agencies has targeted suburban children between the ages of 10 and 12 as a possible interest group to become hunters.[35]

One obstacle to hunting in the early 21st century is the lack of publicly owned land. According to the 2010 census, sixty percent of land in the United States is privately owned. Another 28 percent is owned by the Federal government, mostly in the western states. Nine percent of land in America is owned by the states and 2 percent by Indian tribes. That leaves only 1% of land not owned by anyone. Therefore it is evident that hunters need permission from land owners to hunt on their property. This is not always easy to obtain and is one of the three reasons for the decline in hunting in this country. The other two reasons are scarcity of game in any particular area and high lease costs.[36]

There are a good number of hunters who kill deer in order to eat venison (Latin = venor = to hunt). Others participate in deer hunting only to enjoy the hunt or to bring home a trophy.

Because hunting in urban areas is difficult, those who can afford to pay the costs hunt in Alaska and in foreign countries including Africa. Alaska hunting is best achieved with a bolt action rifle such as the Winchester Model 70, the Browning A, the Remington 700, and a number of others. The cost of hunting in Alaska varies depending on the animal hunted and the payment to so-called "outfitters" and other professional hunters, whose help is generally needed to hunt in the isolation of that state. Nevertheless, additional costs are air fare and clothes, auto expenses, gratuities to guides, and numerous other expenses not foreseen. Therefore, one week bear hunting in Alaska costs about $16,000.

Even more expensive are hunting trips to Africa to hunt leopards and other dangerous animals. The cost of such "safaris" varies with the length of the hunt, beginning with one week or even one month. Airfare alone to Africa is at least $2,000, and the outfit and the need for an experienced guide will easily add up to $10,000 or more.[37]

Expensive hunting is not confined to faraway places. In the United States, as in England, wealthy individuals join elite hunt clubs. This is an overwhelmingly male activity. Some of these hunt clubs were founded as early as the 18th century. In England, the so-called aristocracy has participated in fox hunting while riding a horse, dressed in distinctive hunting clothes and following dogs also called hounds. American imitators of English conduct, mainly members of the exceptionally wealthy "upper class," engage in fox hunting, now banned in England. Because fox are rare in the U.S., coyotes are hunted instead. Coyotes can run forty miles an hour and are therefore capable of outrunning the hounds.[38]

VI.

All governments have the obligation of protecting citizens from outside and inside enemies. Therefore, police and other law enforcers, as well as a defense force, is found everywhere and no less in the United States, where police carry handguns and make use of other weapons. For years, American police carried the six shot .38 caliber revolver (a caliber is the diameter inside a tube or cylinder). This gun has a long tradition because it became associated with "the old West" of legend and movie fame. Movie heroes are often seen spinning the revolver on their forefinger and practicing the "quick draw." Recently, most police departments have abandoned the .38 and acquired the Smith and Wesson .45 caliber semi-automatic pistol. Police believe that "criminals" have greater firepower today than ever before and that they must therefore match that advantage. It is also quite evident that police are intent on impressing the public with their power over the citizenry, so that a larger gun, uniforms, badges, sirens, and other paraphernalia give "the cops" a sense of superiority resting on such equipment.Police guns range in price from about $550 to $670, although a few are cheaper.[39]

Police are taught to shoot at shooting ranges with paper targets. More recently, police shooting training simulators have been used. These simulators create highly realistic situations, allowing the officer to practice and gain immediate feedback concerning his actions. Simulators are safe but nevertheless allow the trainee an opportunity to learn how to make quick decisions in frightening situations. These simulators allow the trainee to learn how to react if a suspect goes for a gun and how to deal with the unexpected.[40]

In numerous communities, police are equipped with "Tasers." "Taser" is a trade name for a stun gun, shaped like any pistol, which fires electrified darts up to 21 feet. The charge shocks the victim, leading to extreme pain, disorientation, heart disturbance including cardiac arrest, and can cause death. The stun guns are also sold to anyone willing to pay between $300 and $800. Contrary to police claims, more than 110 people are killed by police Tasers every year. Tasers deliver a 50,000 volt shock to the victim in a 5 second cycle. The shock renders the victim instantly immobile unless the user holds the trigger down longer in which case the Taser kills.[41]

There are numerous examples of police using Tasers in a reckless and dangerous manner. In Saratoga Springs, New York, an officer tased two victims twice. Both were drunk and fighting. Another victim of a taser shooting was a "bipolar" man who ran through the streets naked and bleeding.[42]

On August 7, 2011, a University of Cincinnati campus police officer shot a high school student with a Taser. The 18 year old student, Everett Howard died of cardiac arrest. The occasion of this senseless killing was that Howard was seen walking outside a university dormitory with balled fists.[43]

In Lindenhurst, Long Island, New York, the police killed Daniel McDonnell by tasing him twice because "he attempted to harm himself" in a police cell. McDonnell was arrested and jailed because of a minor dispute with a neighbor.[44]

In California, Tasers are widely used by police as deaths by Taser have become numerous. In San Jose, California, Richard Lua was killed by police Taser in front of his own garage during a "routine patrol." Lua was the sixth death by Taser in San Jose between 2005 and 2009.[45]

In Onondaga County, New York, Audria Harmon was tased twice by police officer Sean Andrews after a traffic stop while her two children were in her car. There was no legal reason for the assault on Mrs. Harmon. Andrews therefore invented four charges, all of which were dismissed by the district attorney. Andrews shoved Harmon into the back of his car and tased her while handcuffed.[46]

Innumerable additional incidents of police tasing citizens for minor conduct deviations or for no reason at all can be found in all the media every day.

VII.

The use of firearms is the very core of military endeavors. Therefore the American military, like all others, has used guns since the day in 1775 when at the Battle of Breed's Hill and Bunker Hill, the war for independence first began. The American forces used the muzzle loaded flintlock musket and some rifles. A musket has no rifling to spin the ball but is smooth bored. The musket used black powder and held a bayonet. The loading of these muskets was most cumbersome and slow. This was of course not understood at the time, as no one had ever anticipated the modern rifle. Soldiers were expected to shoot every 15 seconds for four minutes. Then the soldier took a rest to recover from the foul odor of the barrel.[47]

Numerous battles against the Native Americans or Indians were fought by American armed forces, who also operated in China, the Barbary Coast, and in several South American countries.

A major war was the War of 1812 against the British, which was fought with the 1795 Springfield musket, copied from the French 1766 model. This musket was named after the first U.S. armory, located in Springfield, Massachusetts.

The Mexican War of 1846-1848 was fought with rifles and muskets. Pistols and Colt revolvers were also used. The standard musket was the .69 caliber smoothbore flintlock musket, with an effective range of about 100 yards. Ten different models were used during the war, with the most famous being the 1822 model. Some troops carried the half-breech-loading flintlock rifle or the 1841 percussion musket. In addition, some officers used the double barreled shotgun in close combat. Cavalry carried a breech loading Hall carbine with a short barrel.[48]

The Civil War (1861-1865) involved far more advanced guns with greater killing power than anything used before. Most important was the invention of the Gatling gun, invented by Dr. Richard Jordan Gatling. This was a hand crank operated weapon which fired 600 rounds a minute out of 6 barrels, each barrel firing 100 rounds a minute.

Handguns, rifles, muskets, breech loaders and repeating guns and grenades were also used by both the Union and Confederate armies.[49]

In the Spanish-American war of 1898, the United States Navy used the Colt machine gun for the first time. That gun was mounted on a light landing carriage which also carried eight ammunition boxes. Developed by John Browning, it was rejected by the U.S. Army, which used the Gatling gun and also used the Winchester-Lee 1895 straight pull action rifle.[50]

The United States entered the First World War in 1917 after it had already lasted three years in Europe. By then the Colt Browning machine gun of 1890 was considered obsolete, although the Italian army used it. The French army used the Hotchkiss 1914 model machine gun, which had been developed in the U.S.A.The Lewis Automatic machine gun was used by the United States Army in World War One. This gun was nicknamed "The Belgian Rattle Snake."The gun was invented by the American Colonel Isaac Newton Lewis in 1911. The gun was widely used by the Allies, but only in a limited manner by the United States. American forces in 1917 relied heavily on the Hotchkiss machine gun, which had only 32 parts and no screws. It was easier to carry than any previous machine gun and fired from an open bolt.[51]

In World War II, the United States army was mainly equipped with the M1 Gerand rifle and the M1 Gerand carbine. In addition, the Springfield 1903 was still in use, as was the 1917 Enfield rifle. Several types of shotguns were also in use, as were anti-tank guns, flame throwers, grenades, mortars, and knives.

Three kinds of handguns were in use. These were the Colt .38, the Colt .45 revolver, and the Colt .45 automatic.The Browning .50 caliber machine gun and the M1919 machine gun, together with the Browning automatic rifle and the Thompson submachine gun, were used by some units in the Second World War.

This equipment meant that American soldiers had five times more firepower in the Second World War than in the First World War. The infantry in the Second World War also carried its own anti-tank weapons in the form of the 3.5 inch Bazooka, named that because of the sound it makes.[52]

The Korean War, from 1950-1953, was mainly fought with Second World War equipment. In use was the M1 rifle, the caliber .30 carbine, the Browning automatic rifle, the .30 caliber machine gun, and the large caliber .50 machine gun. Hand grenades were also used in a variety of forms, as was the bazooka. This equipment was less than adequate in face of the Chinese army and its equipment. The reason for this lack of preparedness by the U.S. armed forces was the belief within the Truman administration that conventional warfare had ceased forever in view of the atomic bomb.[53]

The Viet Nam war lasted ten years, beginning in 1955 when some Americans were sent to Viet Nam as advisors to the South Viet Nam military. The war ended in 1975 after the North Viet Nam forces overran Saigon and American units left that country. During that war, the U.S. armed services were mainly equipped with the newly developed M14 rifle, which was based on the World War Two Garand M1. The M1 was semiautomatic and had an eight round fire capacity. The M14 became fully automatic and used a detachable clip magazine holding twenty rounds.

This gun had obvious advantages over the old M1 but also had drawbacks. The wood stock swelled in the jungle heat and moisture of Viet Nam, which affected accuracy.[54]

In the Gulf War against Iraq between August 1990 and February 1991, the United States armed forces introduced the Squad Automatic Weapon, which was used in addition to the M16 rifle and other older guns. The SAW is a lightweight gas operated magazine that fires from the open bolt position. It can be employed as a machine gun. It can use M16 ammunition. The SAW has become the standard weapon of the infantry.[55]

The most recent war of the United States against Iraq and Afghanistan has led to the development of the XM25 rifle. This gun fires radio controlled "smart" bullets which are programmed to explode when they have traveled a set distance. This allows the shooter to target enemies no matter where they may hide. This rifle has a range of 2,300 feet, so that the shooter is well out of reach of all conventional weapons. This new rifle has a laser gun sight rangefinder which determines the exact distance to the target.[56]

The U.S. armed forces also use older weapons, including the AK47 assault rifle and the M4 also used by terrorist forces.

The gun culture in the U.S.A. is fueled by many sources. These sources include the Second Amendment to the Constitution, the National Rifle Association, guns used in warfare, the ownership of guns in private hands, the gun industry, hunting, and the police. Consequently, guns have become a source of entertainment in the United States, as visible in the endless media events involving guns and violence.

VIII.

Violence by gun is the most common form of violence shown on television, in video games, and in popular movies. For example, Gun Crazy depicts a woman who says "I can only kill" and "Guns make me feel good" as she demonstrates these views in a most brutal manner. Then there is Rambo, in which the actor

Sylvester Stallone kills and kills and kills with guns roaring in a virtual bloodbath. That movie has no meaningful plot, as few movies do, but evidently entertains an audience happy to see as much mayhem as the producers can deliver. Perhaps the most revolting of all of these gun toting movies is Kill Bill, which has two parts. This deals with a woman called "the bride" who seeks revenge on a gang of killers who massacred her wedding party and tried to kill her. In one sequence "the bride"is shot full of rock salt with a double barreled shot gun. She is then buried alive. She nevertheless escapes to engage in further horrors, including plucking out a woman's eye and poisoning others. Even one sequence in which a mother watches television with her child depicts them watching Shotgun Assassin. The entire movie is pointless nonsense except for the gun violence, which evidently made this a big moneymaker. Movie goers spent $152,152, 461 on Kill Bill.

Once Upon a Time in Mexico is another example of a vacuous movie without any merit but with plenty of gun violence. Drug dealers are depicted in an effort to shoot the president of Mexico. When this does not succeed, the drug lord kills a mother and child. A former FBI agent comes out of retirement and kills a man as an act of friendship towards a "buddy."This movie earned its producers $98,185,582. Other movies exhibiting similar violence and bloodshed also earned large profits.

The "Parents' Television Council," in a publication called Dying to Entertain, estimates that the most recent television programming includes 4.41 instances of violence per hour during "prime time." The PTC concluded that violence during the 8 p.m. "family hour" increased by 45%, during the 9 p.m. hour by 92% and 167% during the 10 p.m. hourbetween 1998 and 2006.

That study also concluded that violent scenes increasingly include a sexual element, and that the fetishists and predators seen on such programs as Law and Order, C.S.I., Crossing Jordan, Prison Break, House, and many others invariably resort to guns.[57]

Further studies show that the average child, watching two hours of cartoons a day, will see nearly 10,000 violent incidents each year, leading to such children becoming desensitized to violence.[58]

There are numerous studies of the influence of media violence on youth. These studies indicate that video games and the internet, which are interactive, heighten the consequences of media violence on young people. One of the most violent video games is Mortal Kombat. This became a bestselling game ever since it was first produced in 1992. The game deals with fictional characters who save the earth from evil sorcerers. The game is played by first reading the information concerning the characters in the story. This information is displayed on the screen. The players then seek to defeat various opponents in tournament combat by shooting at the fictional evildoers. This nonsense has also been made into a movie series. Furthermore, multiple editions of Mortal Kombat were produced later.

Lethal Enforcer was another video game. This game was produced in Japan and was best known for its revolver shaped gun called a Konami. The story pertaining to this game takes place in Chicago and features a police officer who stops a growing crime wave using a magnum revolver automatic pistol, a shotgun and an assault rifle. The player can lose one or more lives in the course of the game, as shooting is the main attraction.

There are many studies, undertaken by The American Psychological Association, the Federal Trade Commission, the American Academy of Pediatrics, The American Academy of Child and Adolescent Psychiatry, The American Medical Association, The American Psychological Association, The American Academy of Family Physicians, and the American Psychiatric Association, that all agree that the viewing of violence leads to increases in aggression by both children and adults. In fact, these studies have shown that those who became violent as children are likely to become violent as adults.[59]

Media violence increases excitement, which may then transfer into aggression. This is by no means true of everyone or every exposure to media violence. However, those already prone to violence are most likely to act out what they have seen on TV or done in a video game. Men who consistently watch violence on television are considerably more likely to use violence against women than are men who do not watch such violence.[60]

Exposure to violent "rap" music videos is also associated with violent behavior. These videos endorse violent conduct in conflict situations leading to a significant increase in violent behavior because these videos lead to a greater acceptance of violence as a means of conflict resolution.[61]

Video games are even more likely to provoke violent behavior because those who play such games are active participants rather than mere observers. A number of studies have agreed that there is a connection between playing violent video games and the increased chance of aggressive behavior. Violent video games have an effect on both short term and long term behavior in that repeated viewing of such games increases aggression.Such game series as "Resistance" or "Uncharted" or "Palisade Guardian" involve a great deal of shooting with modern weapons as well as other means of inflicting pain and death.[62]

There have been numerous popular songs dealing with guns, such as "Bang Bang My Baby Shot Me Down" sung by Nancy Sinatra. A group who calls itself "Clash" recorded a song called "Tommy Gun" in 1978. Elton John, an English singer, wrote "My Father's Gun," which is a song of revenge by a southerner whose father was killed in the Civil War. Cheryl Wheeler sings, "Don't forget the guns," which describes a vacation of a family with children who are packing their car. Here the chorus advises the family to bring their pistol. A band who call themselves "Judas Priest" sang "All Guns Blazing." This song of violence is a prime example of gun culture music.

Although the grammar used in these lyrics is most dubious, the message is clear. Violence as the end of a gun is "fun" and entertaining. Evidently the writers of such lines know nothing about the outcome of real shootings and real human suffering.

Summary

There has been an ongoing dispute concerning the meaning of the 2nd Amendment to the Constitution of the United States. This dispute has been largely settled in favor of gun ownership by the American public, especially as there has been a sharp decline in the rate of homicide in the U.S. during the decade ending in 2010. Nevertheless, youth gangs with guns have led to a number of laws regarding gun ownership, defended most vociferously by the National Rifle Association. War, hunting, and the use of guns and Tasers by the police contribute to the gun culture in America as do movies, television, video games, and other entertainment.

Notes

1. Second Amendment to the Constitution of the United States as ratified December 15, 1791.

2. 554 U.S. 570 (2008).

3. H. Richard Uviller and William G. Merkel, *The Militia and the Right to Arms or How the Second Amendment Fell Silent,* (Durham, NC: Duke University Press, 2002), p. 227.

4. David Hemenway, "The U.S. Gun Stock," *Injury Prevention,* vol. 13, (2006), pp. 13-19.

5. U.S. Department of Justice, Federal Bureau of Investigation, *Uniform Crime Report,* (Washington, DC: 2009) Supplement to Return A, no page.

6. U.S. Department of Commerce, U.S. Census Bureau, (Washington, DC: U.S. Census 2010).

7. U.S. Census Bureau, (Washington DC: *Statistical Abstracts of the United States,* 2011), p. 65.

8. Edward Shine, "The Junk Gun Predicament: Answers Do Exist," *Arizona State Law Journal,* vol. 30, (1998), pp. 1183-1207.

9. United States Government, Gun Control Act of 1968, 18 U.S.C. Chapter 44.

10. Christophe Birbeck, et. al. "Controlling New Mexico Juveniles' Possession and use of Firearms," *Justice Research and Policy,* vol. 1, (1999), pp. 25-49.

11. U.S. Congress, 73rd Congress, Session II, Chapter 757.

12. No author, *American Academy of Pediatrics,* vol. 79, (1987), pp. 473-474.

13. Philip J. Cook, S. Moliconi, and T.B. Cole, "Regulating Gun Markets," *Journal of Criminal Law and Criminology,* vol. 86, no.1, (1995), pp. 59-92.

14. Ibid., p. 65.

15. Franklin E. Zimring, "The Medium is the Message: Firearm Caliber as a Determinant of Death from Assault," *Journal of Legal Studies,* vol.1, no. 1. (1972), pp. 97-123.

16. Garen J. Wintermute, "Relationship Between Illegal Use of Handguns and Handgun Sales Volume," *Journal of the American Medical Association,* vol. 284, no. 5, (2000), pp. 566-567.

17. *International Directory of Company Histories,* (St. James, Missouri, 2001).

18. Stephen Hunter, "Triggering Memories: "At the NRA Museum, Tommy Gun Devotees Can Zero In on a Classic," *The Washington Post,* (March 22, 2004), pp. CO1.

19. Stephen Halbrook, "Nazi Firearms Law and the Disarming of the German Jews," *Arizona Journal of International and Comparative Law*, vol. 17, no. 3 (2000), pp. 483-585.

20. Michael A. Bellesiles, *Arming America: The Origin of a National Gun Culture*, (New York: Alfred Knopf, 2000), p. 371.

21. Arthur Cecil Bining, *A History of the United States*, vol. II, (New York: Charles Scribner's Sons, 1950), pp. 464-474.

22. Paul Fargis, Editor, *The New York Public Library Desk Reference, 3rd edition*, (New York: Macmillan Publishers,) pp. 831-840.

23. Bellesiles, *Arming America*, pp. 377-381.

24. David A. Hounshell, *From the American System to Mass Production: The Development of Manufacturing in the United States*, (Baltimore, MD: The Johns Hopkins University Press, 1984), pp. 18-19;49-50;87-92;332-333.

25. Gideon Burrows, *The No Nonsense Guide to the Arms Trade*, (London: Verso, 2002), pp.111-117.

26. http:www.handgunsforsale.net and http://www.firearmdeals.com

27. Marye Audet, "Antique Gun Collectors," *Love to Know Antiques*, http://antiques.lovetoknow.com/AntiqueGunCollectors

28. Frederick Kunkle, "Gun Enthusiasts Find a Paradise of Wood and Steel," *The Washington Post*, (June 25, 2006), p. CO3.

29. John Holusha, "Mass Killer is Recalled as a Gun-Loving Youth," *The New York Times*, (July 21, 1984), p. 8.

30. Judith Cummings, "Neighbors Term Mass Slayer A Quiet but Hotheaded Loner," *The New York Times*, (July 20, 1984), p. A1.

31. Dale Russakoff, "An Ordinary Boy's Extraordinary Rage," *The Washington Post*, (July 2, 1995), p. AO1.

32. Larry Whiteley, "Nearly Seven Million People Expected to Attend 2011 Fall Hunting Classic Across America and Canada," *PRWeb*, (Springfield, MO, July 20, 2011), pp. 1-2.

33. No author, *Guns America: Where America Buys and Sells Guns*, http://www.gunsamerica.com/

34. Howard J. Kirkpatrick, "A controlled Archery Deer Hunt in a Residential Community," *Wildlife Society Bulletin*, vol. 27, 1999, pp. 115-123.

35. Jodi DiCamillo and Josph M. Schaefer, "Internet Program Impacts Youth Interest in Hunting," *Wildlife Society Bulletin*, vol. 28, no.4, (2000), pp. 1077-1085.

36. Clark E. Adams, Robert D. Brown, Billy J. Higginbotham, "Developing a Strategy Plan for Future Hunting," *Wildlife Society Bulletin*, vol. 32, no. 4, (Winter 2004), p. 1157.

37. No author, "African Hunting Packages," http://www/afrkanskyhunting.co.za/africanhuntingpackages.html

38. Daily Mail Reporter, "Tally-ho on the Range! Fox Hunting Gets the American Treatment," *Mail Online*, (April 11, 2011), p. 1.

39. R.H. Hodges, "Facts and Policy on Police Guns," *The New York Times*, (April 2, 1989), New York/Region, p. 1.

40. Linda Woolheather, "The Advantage of Police Shooting Training Simulator," Money, (October 22, 2010), p. 1.

41. Peter Gorman, "Torture by Taser." *Fort Worth Weekly*, (June 23 2005), p. 3.

42. Dennis Yusko, "Taser used in Spa City detailed," *Times Union (Albany, NY)* July 14, 2011, p. Business News.

43. Jim Hannah, "Student Dies After Police use Taser at University of Cincinnati," *The Cincinnati Enquirer,* August 7, 2011), p. 1.

44. No author, "LI Man dies after Tasering," *New York Post,* (May 7, 2001), p. 14.

45. No author, "Death by Police Taser," *San Jose Mercury,* (February 2, 2009), p. 1.

46. Mike Celizic, "Mom Tasered in Front of Kids Posed No Risk," *Today,* (August 14, 2009), p. 1.

47. George C. Neumann, *The History of Weapons of the Revolutionary War,* (New York: Harper & Row, 1967).

48. George J. W. Urwin, *The United States Infantry,* (New York: Sterling Publishing Co., 1991), p. 66.

49. Dennis Adler, *Guns of the Civil War,* (Minneapolis, MN: Zenith Press, 2011), pp. 97, 155, 193.

50. Trevor K. Plante, "New Glory to Its Already Gallant Record," *Prologue,* vol. 30, no.1, (Spring 1998), p. 1.

51. Hew Strachan, *The Illustrated History of the First World War,* (New York: The Oxford University Press, 2000), pp. 142, 145, 183, 248 and 287.

52. Richard A. Gabriel and Karen S. Metz, *A Short History of War: the Evolution of Warfare and Weapons.* (Carlisle Barracks, PA: Strategic Study Institute, U.S. Army War College, 1992).

53. Ibid., chapter 5.

54. James Wesley Rawles, "Round Capacity Magazine," *The M14/M1A Magazine,* (November 17, 2008), p. 1.

55. No author, "Squad Automatic Weapon (SAW), M249 Light Machine Gun," *Military Analysis Network,* (January 20, 1999), pp. 1-4.

56. Daily Mail Reporter, "No Hiding Place from New U.S. Army Rifles that Use Radio Controlled Smart Bullets," *Mail Online,* (November 30, 2010), pp. 1-7.

57. Caroline Schulenburg, *Dying to Entertain,* (Los Angeles, The Parents' Television Council, 2006), p. 1.

58. Barbara Wilson, "Violence in Children's Television Programming," *Journal of Communication,* vol. 52, (2002), pp. 5-35.

59. Richard E. Tremblay, "The Development of Aggressive Behavior During Childhood," *International Journal of Behavioral Development,* vol. 24, (2000), pp. 129-141.

60. Len Berkowitz and J. T. Alioto, "The Meaning of an Observed Event as a Determinant of its Aggressive Consequences," *Journal of Personality and Social Psychology,* vol. 28, (1973), pp. 206-217.

61. James D. Johnson, J. T. Jackson and L. Gato, "Violent Attitudes and Deferred Academic Aspirations: Deleterious Effects of Exposure to Rap Music," *Basic and Applied Social Psychology,* vol.16, no.1-2 (1995), pp. 27-41.

62. Jeanne B. Funk, G. Flores and D. D. Buchman, "Rating Electronic Games: Violence is in the Eye of the Beholder," *Youth and Society,* vol. 30, (1999), 283-312.

Chapter Six

The Telegraph and the Telephone

I.

Samuel Morse (1791-1872) is commonly credited with inventing the telegraph. Yet neither he nor any one person can be credited with it. The fact is that numerous men contributed to that invention, although the Morse code is undoubtedly his crowning achievement.

The word "telegraph" means distance writing. While that was not truly achieved until the 19th century, even the most ancient civilizations attempted to communicate at a distance. An excellent example was the use of the ram's horn, or shofar, by the ancient Israelites, who used that horn to signal from one mountaintop to another. Other ancient peoples used fires lit at intervals on mountaintops. Early men also knew that electricity and magnetism existed, but were unable to explain either phenomenon or to make use of them.[1]

It has been suggested that the Olmec who lived in Central America may have discovered the geomagnetic lodestone compass before 1000 B.C., predating the Chinese discovery by more than a millennium.[2]

The existence of electricity was of course known long before it was put to any human use. Lightning was everywhere, and some fishermen knew that there were electric eels and electric catfish.

The Chinese scientist Shen Kuo (1031-1095) was first in writing about the magnetic needle compass, which improved navigation by employing "true north." In thirteenth century France, Peter Peregrinus wrote a treatise on magnetism, and in the fourteenth century, the Italian Flavio Gioja invented the dry compass.[3]

A number of other inventors contributed to human knowledge about electricity. Among these, the Englishman Dr. Robert Gilbert is credited with naming the phenomenon "electricus" from the Greek word for amber. These and many other contributors became the forerunners of the inventions that led to the telegraph. Included in this array of scientists are LeMonier, Watson, Franklin,

Lesorge, Lomond, Cavallo, Salva, Oersted, Ampere, Schilling, Gauss, Weber, Wheatsmore, Henry, and finally Morse and his associate O'Reilly.[4]

The telegraph was the first digital technology, in that it transmits information by turning an electric circuit on and off. This made possible instant communication over long distances, thereby changing the basic social institutions, much like printing had done in 1440 when the German Johannes Gutenberg used that Chinese invention to print a Bible.

Telecommunication became a major force in the nineteenth century in the United States. This can be understood by comparing the speed of communication before and after the invention of the telegraph. For example, before 1840 when railroads became available, it cost 25 cents to send a letter from New York to Chicago, a distance of 850 miles. By 1850, the railroads had reduced that to two days at a cost of 3 cents. During that same decade, 1850-1859, the telegraph reduced the speed of sending one page from New York to Chicago for $7.50.[5]

In 1890, when railroads were in the ascendancy, their value was about $8 billion, while the telegraph industry was valued at no more than $147 million. In that same year, employment in the telegraph industry was around 50,000, but railroad employment was at least 500,000.[6]

It was in 1600 that William Gilbert published "On the Magnet, Magnetic Bodies and the Great Magnets of the Earth," in which he used the terms "magnetic pole," "electric attraction," and "electric force," implying some kind of magic allowing people miles apart to communicate with one another. These stories were believed because lodestones or magnetized pieces of rock attract iron. Such stones have been found in the United States and elsewhere. These stones were used in earlier centuries as magnetized needles.[7]

Beginning in 1837, numerous individuals demonstrated the telegraph by using static electricity, with the result that the first patent for an electromagnetic telegraph was issued to William Cooke in England in 1837.By that time, the portrait painter Samuel Morse had already become interested in the telegraph, because he met Charles Jackson, a student of electromagnetism, on board a ship sailing to the United States in 1832. Morse was fascinated by the speed of transmission along electric wires and therefore experimented with electromagnetism until he was able to develop his telegraph and the "Morse code" in 1838. Then, in 1845, Morse founded the Magnetic Telegraph Co.[8]

Morse had been a portrait painter with some success. Nevertheless, he renounced that profession and gave up painting entirely, because he believed he had failed as an artist. In 1982, more than a century after his death, his painting Gallery at the Louvre was sold for $3.25 million, which, at that time, was the highest amount paid for a painting by an American.

Because Morse viewed himself as a failure as an artist, he turned to another endeavor. It is no idle speculation to say that if Morse had not believed he had failed as an artist he would not have invented the "distance writer." It was that "failure" which led Samuel Morse to pay attention to developing the telegraph. That device consisted of a system of electromagnetic relays, so that it put no limit to the distance a message could be sent. Added to this invention was the Morse code, which Morse worked out by devising a system in which the alphabet was transmitted in dots and dashes.[9]

Morse demonstrated his invention in Paris, France in 1838. This led to his widespread recognition, a role he had craved as a painter without success. Morse was acclaimed in Europe and America, so that by 1844 Morse had built a telegraph line between Washington and Baltimore with Congressional approval. Thereafter, the invention of the telegraph was widely accepted, so that two years after Morse had demonstrated its usefulness, there were nine telegraph companies in the United States, with 2000 miles of wire stretching from Maine to New York and from Cincinnati to Chicago and to New Orleans.[10]

Morse did not only seek to publicize his invention in the United States. He and his associates demonstrated its use in Europe and in Asia. To do this, Morse approached wealthy people and governments in an effort to gain support. For example, in 1847 Morse explained his telegraph to the Sultan of Turkey, and later took it as far as Japan.[11]

II.

Prior to the invention of the telegraph, there was no standard time in this country. Then, the United States Naval Observatory and Western Union cooperated in creating this development, so that ever since then, Americans have been "tied to the clock."This dependence on the clock began with the need to standardize time for factories, which used time clocks to assemble the work force and to calculate pay by the hour. During the rural era in the United States, which gradually disappeared in the middle of the twentieth century, sunrise and sunset were enough to delineate the workday. In factories, however, it became necessary for the owner to know who worked in the several shifts dividing the workday. Control of time is of course a major part of capitalist life, and that was made possible by the clock, which was in turn dependent on the telegraph.[12]

When rural workers first came to work in American factories, they viewed these factories as a kind of jail because management controlled the shift clocks, which were not coordinated from one location to another and which employees did not understand because they themselves had no watch or clock. This in turn led to the widespread acquisition of pocket watches, and later the wristwatch.[13]

In the nineteenth century, railroads provided the fastest means of communication in human history. As the country expanded and eventually reached the west coast,railroads followed. Therefore the telegraph became most important after rail travel from the east coast to the west coast became available in 1869. Prior to the invention of the telegraph, railroad time was set according to the nearest large city. This resulted in having two clocks at each station, one for local time and the other for the line time. This dilemma was finally resolved when the telegraph could be used to know what time it was all along the line in every area of the country. This then led to the invention of standard time and the segregation of time zones across the country in 1883, and for the whole world in

1884 at the Meridian Conference in Greenwich, England, and to the invention of Greenwich Mean Time used by all nations today. This is an excellent example of unanticipated consequences, resulting from the invention of the telegraph.[14]

III.

Because the telegraph made prompt communication over a great distance possible, it also had major implications for international relations in peace and in war. The Anglo-French telegraph cable was first installed in 1851.This and subsequent international telecommunications remained in private hands because governments were not willing to let foreign nations to operate on their soil. Moreover, the decade of the 1850's saw the introduction of the first transatlantic cables, which were primarily owned by British companies. The cost of laying such cables was enormous, so that the British government funded a good part of the expense of doing so.[15]

Nevertheless, the cost was worth it for the British government, as the British had the largest commercial navy and well as a large military navy. The telegraph cable made it possible for sea captains to stay in touch with their home bases, which was quite impossible until the telegraphic cable was invented. Prior to the invention of the telegraph, a ship usually disappeared for months if not years.

Earlier there had been several unsuccessful attempts to lay submarine cables, but it wasn't until Thomas Crampton succeeded in laying a cable across the Channel in 1851 that a new industry was launched. The cables were laid by special cable ships, which were equipped with storage tanks for cables. These ships laid cables for military and political uses. In 1857, a transatlantic cable succeeded in allowing U.S. President Buchanan to contact Queen Victoria, although it took 16 hours to deliver the messages. By 1885, all the continents had been connected to cable, which revolutionized communications around the world. The time needed to deliver a message by cable gradually lessened, so that it was almost instantaneous when the First World War erupted. During that war, cable was first used to keep in touch with the American Expeditionary Force in Europe.

Beginning at the end of the 19th century, governments used cables to keep connected to their diplomats overseas and elsewhere. The British held a near monopoly on transatlantic cabling until Germany was able to connect to the United States by means of cable in 1900.

There was therefore a great deal of telegraphic interaction provoked by governments, including the infamous "Zimmermann telegram" ofJanuary 16, 1917, which had at least some influence on President Woodrow Wilson in his decision to enter the First World War on the side of the Allies and against Germany.

Arthur Zimmermann (1864-1940) was the State Secretary for Foreign Affairs of the German Empire in 1917. He sent a coded message by telegraph to the German ambassador in the United States, Count Johann von Bernstorff, on January 16, 1917, with the request that the contents be forwarded by telegram to the Geman ambassador in Mexico, Heinrich von Eckardt. The telegram told Eckardt that in the event the United States appeared likely to enter the European war on the side of the Allies, that Eckhardt should provoke the Mexican government to attack the United States. The German government offered Mexico a military alliance and promised to furnish Mexico with military assistance in the form of weapons. Mexico was to regain all the territory it had lost to the United States, including Arizona, New Mexico, and Texas.The idea was to keep the United States so busy fighting Mexico that the United States could not simultaneously fight Germany.

The telegram consists of seventeen rows of numbers, constituting a code which the German government believed would hide the contents from anyone who had no access to their code book. The Germans did not know that the British had deciphered the code and were therefore in a position to deliver its contents to President Woodrow Wilson. Subsequently, the contents of the telegram were published in American newspapers on March 1, 1917. The United States declared war on Germany on April 6, 1917.[16]

The written telegram made diplomacy around the world more transparent than it was in previous times, when diplomacy was carried out by verbal discussions between ambassadors or other politicians. The speed of information increased, leading to more emotional responses to the contents of telegraphed messages and therefore more haphazard decisions. Traditionally, diplomats were upper class snobs who prided themselves on not having to work. As the telegraph increased the speed and the volume of diplomatic communications, more and more decisions were made in haste and with poor judgment. The earlier reliance of "envoys" was greatly diminished by the telegraph and led to the promotion of ever more sophisticated surveillance techniques.[17]

When the transatlantic telegraph was established in 1866, many commentators believed that this new device would ensure world peace, as the leaders of countries far apart could now communicate with one another rapidly and settle their grievances. Unfortunately, the long-sought peace did not happen. Instead, more wars were fought since then.

The true beneficiaries of the telegraph were not the governments but the stock market, the railroads, and the newspapers. The newspapers began to rely on the Associated Press, which was the outgrowth of the telegraph, as three news organizations were able to merge into the AP although located in various regions of the country. This led to the employment of women as telegraph operators, a development foreshadowing the gender revolution to come.[18]

The railroad companies also found a great use for the telegraph. The telegraph allowed them to coordinate schedules and to avoid train wrecks. The telegraph allowed railroads to ensure more cooperation between units distant from each other and to create a nationwide single tracked railroad system.

The telegraph improved trading on the stock exchange. It lowered prices, increased rapid execution, and led to confirmation in one and one half minutes.

By 1909 it took only three minutes to execute a trade across the Atlantic in London. By 1920 it took less than 56 seconds to deal with a trade between a San Francisco brokerage house and brokerage in New York City.[19]

After the telephone became in common use around 1893, the telegraph business declined. This led Western Union to sell "social telegrams," which resembled greeting cards in that they were impersonal but nevertheless effective in greeting friends and relatives on birthdays, weddings, funerals and holidays. In fact, the telegraph finally had to compete with numerous other devices, so that wiring money became the sole source of income for the former telegraph industry.[20]

IV.

It is by no means an exaggeration to view the telegraph as a major contribution to "modern society." The telegraph created dependence on time for nearly everyone and contributed to the centralization of control, leading to major productivity growth in almost all industries, and revolutionized the conduct of business.

This was most dramatically evident in the 19th century, when the telegraph first came into use. At that time, telegraph operations were relatively primitive, as the wire services competed with one another with minimal manpower and feeble resources. As a result, these early telegraph proprietors were often guilty of sending false "news" or coloring news stories to gain more attention and profits. It was an era of high costs and low newspaper circulation, which led to the undue influence of "big business" on the wire services, who were induced to withhold reports unfavorable to business interests or make a great "to do" of news favorable to their investments.[21]

These schemes did not go unnoticed, so that it was commonly believed that the wire services were "watering down" the news in the interests of government, business, or themselves. In theory, the wire services were to be the "great levelers" in the world of finance. Nevertheless, it appeared that large banking houses like the Morgans and their European counterparts were given advance information about stock market transactions and other business related news which these large companies used to drive all others out of business.[22]

The Associated Press and the United Press were the most fierce competitors in the 1890's, both depending of the wire services to maintain their business. It was newsgathering for profit, with the Associated Press winning the fight, as the United Press went out of business in 1895. That press was accused of numerous news suppressions about radical or labor causes. The Associated Press also refused to report on the investigations of Upton Sinclair into the meat packing industry, and their agent in Denver refused to carry Sinclair's telegram to President Woodrow Wilson showing that the government of Colorado misrepresented the state's involvement in a miners' strike.[23]

The wire services generally "killed" damaging reports that affected domestic customers and particularly those that affected overseas investors.

Nevertheless, the telegraph or wire services survived until the telephone challenged their dominance. Before that, little was known of the manner in which wire services gathered the news. The boards of directors made it a point not to answer charges of fraud and greed, because that tactic worked for them in that the charges became "old hat" and were soon forgotten. Furthermore, the newspapers, which depended on the telegraph services, had little control of the telegraph companies. The shares in those companies were almost entirely owned by the board of directors.[24]

In the course of the 19th century, a fierce competition developed between Western Union and the Bell Telephone Co. Because both companies wanted to end this costly rivalry, they signed a contract in 1879 which was intended to partition the electrical communications industry of the United States. This meant that the far larger and more powerful Western Union withdrew from the telephone market and Bell agreed not to compete with Western Union for telegraph business.[25]

Evidently, the Western Union management failed to recognize that the telephone would supersede the telegraph in short order. As a result, Bell grew immensely, while Western Union declined as the telephone more and more replaced the telegraph as the most popular form of communication in America. The Western Union management saw the telephone as a passing fad and did not understand the major opportunity Bell's invention provided. The management of Western Union did not recognize the value of the telephone and was anxious to maintain its monopoly of the telegraph industry. Furthermore, the directors of Western Union feared a takeover by the super-rich Jay Gould.[26]

Despite the agreement of 1879, the Bell Telephone Co. aggressively competed for telegraph customers and succeeded in driving Western Union nearly out of business when Bell's successor company, American Telephone and Telegraph, acquired control of Western Union in 1909. Thereafter, Western Union declined even further, as the depression of the 1930's reduced its business. Western Union was saved from total collapse only because the Federal Communications Commission (FCC) enforced the anti-monopoly provisions of the law. The FCC was also instrumental in forcing AT&T to sell its TWX (teletypewriter exchange) service to Western Union in 1971.[27]

The telephone did indeed supersede the telegraph. This happened not only because the telephone is technologically superior, but because a good deal of legislation created the advantages of the telephone over the telegraph. In addition, a telephone user has more control over the interaction between himself and the party on the other end than is true of someone using a telegraph. Telephones are in the house, and in the 21st century, in one's pocket. Telegraphs were public and provided limited interaction. Even before the widespread use of the cell phone, Western Union became largely obsolete, so that its chairman, Robert Flanagan, described his efforts to save his company in 1985 as trying "to turn an elephant around in a bathtub."[28]

The cable telegram and the entire telegraph industry began to decline after Guglielmo Marconi invented the radio, which made the telegraph obsolete.

Marconi used a beam system short wave radio in addition to the long wave radio he had also invented. This led to the loss of half of the cable companies' traffic and even more when in 1929 an underwater earthquake destroyed numerous cables off the coast of Nova Scotia. Finally, the great depression of the 1930's caused cable telegraph companies to lose so much money that many were forced to go out of business. Consequently, the telephone and live television dominated the communication industry, leaving telegraphs a minimum of customers seeking to send money over some distance.[29]

V.

It is common knowledge that Alexander Graham Bell invented the telephone. This belief is only partially true. Like all inventors, Bell had to rely on a good deal of work done by others before him. Included in these numerous pioneers were Innocenzo Manzetti of Italy, Charles Bourseul of France, and at least ten others.[30]Johann Phillipp Reis of Germany, however, actually invented a telephone in October 1861, fifteen years before Bell presented his invention. The Reis telephone did not always succeed in transferring speech, although it did so sporadically. In short, Reis' telephone was unreliable but nevertheless preceded the work of Bell. As a result, Bell has been given the principal credit for this important invention, which has served mankind ever since.[31]

Bell came to America from Scotland, where he was born on March 3, 1847, when he was twenty-four years old.He arrived in Boston that year to demonstrate a speech system designed to train the deaf. This had been developed by his father. Once in the United States, Bell encountered the effort of numerous men in attempting to use the telegraph lines for talking. Bell began by transmitting musical tones by means of electricity. This had already succeeded when others tried it and it led Bell to attempt sending more than one message over one Morse wire at the same time. Bell then studied electricity and visited a number of laboratories until, by accident, he found a means to send a spoken word over a wire.[32]

On the same day in 1876 on which Bell filed for a patent for the telephone, Elisha Gray also filed for a patent on the telephone. This "coincidence" has led to much speculation concerning the true inventor. The fact is that the telephone was invented separately by both men. Yet, charges of fraud were promoted by both men, as each accused the other of stealing his ideas. These charges led to congressional investigations and much litigation without any conclusive results.[33]

Within about two years after Bell filed for his patent, there were about 10,000 telephones in the United States. By 1893 this number had risen to266,000. Thereafter, the number of telephones reached millions. By 1900, there were 1,356,000 telephones in use. By 1910 the number had reached 7,635,000 and by 1920 it had reached 13,273,000. In 1930 there were 20,103,000 telephones in the U.S.A and by 1980 there were 180 million telephones in use in America. In 2011, 327 million telephones were in use in the U.S.A.[34]

At first, every telephone customer was provided with an individual line from which he could make unlimited calls for about $4.66 a month, which represented about 13% of an average American worker's pay.[35]

That was very expensive, so that it was considered foolish and wasteful to discuss anything but business during a telephone conversation. Idle gossip and entertaining calls were viewed with suspicion, as businessmen used the phone to reach customers or keep in touch with the office when absent. Subsequently the telephone was advertised as a means of household management, such as calling a plumber or inviting someone to dinner. This led eventually to sociability, which was not contemplated by early owners. This changed with long distance marketing, when sociability was used to lure customers to call long distance and spend a good deal more money.[36]

Other strategies bringing in more telephone users was to claim that job offers can come by telephone and that telephones are most useful in cases of emergency. Friendship was also advertised as a possible product of telephone use, as was the admonition to "call the folks."[37]

Farmers' wives were also targeted in order to increase business. "A blessing for the farmer's wife. It relieves the monotony of life. She cannot become lonesome with the Bell service," said one advertisement.[38]

In 1903, The Chicago Daily Tribune published a story concerning the installation of a telephone system in the farm country of upper Michigan which cost $3 a year. This telephone service originated in a wire five miles long, which then started "a telephone craze." This "craze" consisted of telephone meetings in local schoolhouses attended by men, women, and children who discussed how a doctor could now be called if a telephone would be available and how farmers could talk to markets in Chicago or Detroit.

Farmers formed stock companies which set up telephone poles, bought wire and telephones. The cost was about $20 for each participant, as several farmers cooperated in serving several families. General stores then began to take orders over the telephone, as numerous other benefits resulted from the use of the phone. Each family was given a different number of long and short rings. Nevertheless, everyone listened in on his neighbor and enjoyed the ensuing gossip.[39]

As more and more people installed telephones, it became usual to connect two or more customers to one telephone line, which was commonly called a "party line." This permitted people to listen in to other people's telephone conversations. Contrary to the expectation that the working class, those not in business, had no use for the telephone, the so-called "idle" use for gossip and general conversation attracted a good number of users who had no business to conduct.

The telephone made distances more accessible than ever before. Most dramatic evidence of this came in 1915 when Alexander Bell made the first U.S. coast to coast telephone call. Bell called from New York City and his assistant,

Thomas Augustus Watson, answered in San Francisco. Thereafter, telephones became ever more popular until, in the early 21st century, about 5 billion people worldwide own a telephone.[40]

The consequences of this wide use of the telephone have been considerable. Not only is instant communication possible, but in addition, telephones have had widespread commercial uses. Included are public opinion polls, which are conducted almost entirely by telephone.

Survey research is also a product of the telephone. That is mainly about the collection of data concerning the entire American population by using a sample deemed to reflect the answers of the entire population, were they to be given an opportunity to participate.

As early as 1916, The Literary Digest, a popular magazine founded in 1890, conducted telephone surveys and opinion polls leading to predicting the outcomes of presidential elections. The editors continued these telephone polls in 1920, 1924, 1928, 1932, and 1936. The magazine sent out "ballots," also including a subscription blank. These "ballots" were postcards which were then tabulated, correctly predicting the outcome until 1936. In that year a telephone poll was conducted, resulting in the prediction that the Republican candidate, Alf Landon, would win the presidency in a "landslide." According to that poll, the Democrat Roosevelt was to lose in the Electoral College. In fact, Roosevelt received 523 electoral votes and Landon only eight.[41]

The magazine actually sent out 10 million straw ballots, derived from telephone books, resulting in a return of twenty-five percent. The magazine then published the results of the popular straw vote to the effect that Landon received 1,293,669 votes or 55% and Roosevelt 972,897 or 41%, with Lemke, a third party candidate, the remainder. Actually, Roosevelt received 61% of the vote.

The reason for the failure to accurately predict the election was first to overlook that Roosevelt was heavily supported by the poor, who had no telephone. Furthermore, the response rate was less among the poor than among the well-to-do. This problem was exacerbated by the fact that Republicans tended to be better educated voters who were, for that reason, more likely to return a "straw" ballot.[42]

Although the Literary Digest survey of 1936 was a spectacular failure, the telephone has become the dominant method of doing survey research and data collection. This is a major change from practices adopted in the 1950's. The reasons for using the telephone are that almost everyone in America has a telephone, it is the cheapest way of reaching people living at a long distance, sampling techniques have improved immensely since the 1930's, and telephone interviews can be assisted by the use of computers.[43]

The invention of the telephone was at first used by businesses that preferred talking to people over long distances rather than using the telegraph. At home, the telephone was first used by the wealthy. Later, public telephones were introduced, which made telephones available to nearly everyone.

The advent of the cell phone has made public opinion surveys insecure once more. As the use of cell phones increases, telephone accuracy becomes more and more questionable, as there are numerous households as of 2012 who have no land-line telephone but rely on the cell phone alone. This means that telephone surveys cannot reach everyone equally, either because the surveys leave out cell phones, the cell phones are more unlikely to be engaged in an activity (such as driving) that will compel them to forego answering survey questions, or because those with multiple lines will be overrepresented. At the end of 2011, at least 26% of Americans had only a cell phone, while among young people less than thirty years of age, the proportion of users without land line phones was much higher. Therefore, surveys concerning the presidential election of 2012 are insecure at best.[44]

As in the past, it can be assumed that the poorer households are more likely to lack a telephone and may therefore be overlooked by surveyors. Furthermore, telephone surveys are based only on the random use of telephone numbers and therefore lack the any other contact with the respondents on the other end of the line. This means that, unlike face to face contacts, the telephone survey is utterly impersonal and therefore easily interrupted and broken off. Because face to face contact is missing, the telephone surveyor has no means of presenting himself as an authoritative person or one who can be trusted or one who is competent. Consequently, many people refuse to participate in telephone surveys before even hearing any reason for asking for their cooperation.

A ringing telephone will usually be answered without the respondent having an opportunity to identify the caller. Therefore, the respondent will assume that the introductory words of the call will immediately be followed by some request which can easily be interpreted as an unsolicited invasion of privacy.[45]

Telephone surveys are more easily terminated than face to face interviews. In the latter circumstances, the interviewer is usually invited into the respondent's home. Many people will find it appropriate to serve the interviewer coffee, ask him to be seated, and then establish at least a modest interactive relationship to the "other." None of this is possible in a telephone interview, which remains impersonal and distant.

VI.

The most important change in telephone use since its invention has been the cell phone or mobile phone. This development began when, in December 1947, Douglas Ring, an engineer at Bell Laboratories, proposed "hexagonal" cell transmission for mobile phones.

Subsequently, on April 3, 1973, Motorola manager Martin Cooper placed the first handheld cellular phone call to Dr. Joel S. Engel at AT&T's Bell Laboratories.There were more than 327 million "cell phones" in use in the United States by 2010. This means that there are more cell phones is use in the United States than there are people, as the population in 2011 was about 312 million.[46]

This immense use of mobile phones has had a number of consequences for American life at the beginning of the 21st century. This includes the increase in the speed at which information is available and how people meet one another. Among 'teens this is visible in that the old street corner gatherings are no longer practiced. Instead, young people can meet anytime, anywhere, by making arrangements by mobile phone. This ability to meet anyone at anytime is of course also available to adults.[47]

When mobile phones were first introduced to the public, they were sold on the grounds that cell phones provide more security and more safety for their owners. Since then, mobile phones have become so important to so many people that owners have become emotionally attached to them. Many a cell phone user can no longer live without one, although those who have never used a cell phone feel no deprivation. Like so many inventions, consumers are "sold" on the latest technology until they can hardly imagine life without the gadget.

There are those who are so attached to their cell phone that they view the device as an extension of their hand or like a part of their anatomy. Many users literally live by watching their Blackberry light up so they can gossip. Furthermore, many users are of the opinion that "everyone" has a cell phone. This belief leads to the conclusion that one must not be without such a phone, lest one is viewed as an outsider. Lack of a cell phone is therefore seen as a stigma.[48]

The use of cell phones has allowed some people to create a "bubble" around themselves so that they shut out their immediate surroundings as they talk endlessly on the phone. Everywhere there are people talking on their cell phones while walking on the street or driving a car, oblivious to their surroundings and absorbed in their conversation. This kind of telephone use leads to a state sociologists call "civil inattention."[49]

Another consequence of the use of mobile phones is the belief on the part of frequent cell phone users that everyone should always be available on a cell phone. Indeed there are more mobile phones in use in the United States than there are inhabitants. Nevertheless, there are those who do not have nor want to have a mobile phone and who don't want to talk endlessly. Because mobile phones are always at hand, the number of calls made by users increased immensely over the number of calls made from stationary phones. Users of these devices keep friends and relatives informed of their location at all times.[50]

Mobile phone usage also results in changes in travel behavior. There are mobile phone users who coordinate travel plans with others by means of their cell phone. In addition, many a phone user saves himself a trip by using the phone instead. Mobile phones also allow people to make last minute changes in their meetings, such as asking people to meet a little later than planned to accommodate some late event.

Because the mobile phone is also widely used by drivers, its use has been responsible for auto accidents. This is largely true because cell phones are no longer only used to talk to someone, but have also become music players, mobile e mail, internet search engines, digital cameras, and video recorders.

These devices have led to an immense increase in the amount of time spent on talking and using one or the other aspect of the cell phone (if it can still be called that). Thus, a survey showed that 81% of cell phone owners said that they had used a cell phone while driving.[51]

The United States Department of Transportation website on distracted driving reports that annually over one half a million people are injured and over six thousand are killed due to distracted driving. The Department lists three types of distractions that can occur while driving. These are, taking the eyes off the road, taking the hands off the wheel, and taking the mind off driving.[52]

In view of some horrible accidents caused by distracted drivers, numerous state legislatures have considered two hundred bills dealing with this problem. A good number of accidents caused by distracted driving have been caused by "texting," which appears to have become a national obsession, as 173 billion such messages were sent in 2010. Undoubtedly, "texting" is the most dangerous activity while driving, as it involves the sending and receiving of messages and sending electronic mail, reading and writing such mail, or dialing a number. Evidently, text messaging lifts the chances of a crash to 23.2 times the risk of a non-distracted driver.[53]

A study by Professor David L. Strayer of the University of Utah showed that drivers who talk on a cell phone with or without holding the phone are as dangerous as drunk drivers. Holding or speaking on a cell phone while driving is more dangerous than using makeup or eating because it is prolonged over a good deal of time.[54]

Cell phone conversation differs from speaking to a fellow passenger while driving because the passenger is aware of any dangerous situation or special demands on the driver and will therefore stop talking if needed. This is not possible for the respondent on a cell phone.[55]

It is a myth that hands-free cell phones are less dangerous than those held in hand. In fact, two thirds of respondents to a survey believe that hands-free phones are less dangerous than in hand phones. In fact there is no difference in dangerousness because the driver is as much distracted from driving no matter whether he holds the phone or not. The National Highway Traffic Safety Administration has made studies which show that it is the conversation, not the phone itself, which determines that the driver becomes distracted and unable to drive safely while talking.[56]

In view of the widespread use of the cell phone, it is remarkable that baseball clubs continue to use the landline telephone to communicate between the dugout and the bullpen. In the dugouts there are televisions which help the managers to watch the action in the team's bullpen as to who is warming up and who is ready to pitch. Some teams also had specifically designed mobile phones until the phone bills reached truly astronomical proportions. Therefore, most teams have continued to rely on landline telephones. This reliance on the traditional telephone could have cost the St. Louis Cardinals the World Series when they lost game five to the Texas Rangers on October 24, 2011. According to the St. Louis Cardinals manager, Tony La Russa, the Cardinals bullpen coach,

Derek Lilliquist, who was talking on an unmovable landline in a extremely loud area of the Texas ballpark, twice misunderstood as to which pitcher should warm up. As a result the wrong pitcher was in the game at the wrong time, leading to a 4-2 victory for the Rangers.

Regular cell phones are not allowed in the dugouts because they have outside calling capabilities.[57]

In 2011, a crowd of mostly unemployed young adults accumulated in a New York City park displaying signs denouncing "the rich," "the banks," and all Jews. Claiming that they represent the 99% of Americans as against the one percent who reputedly have "all the money," the campers who "occupied Wall Street" were called together and remained largely together by using the cell phone. Inspired by the canards of the past, those who participated in these gatherings on the streets of a number of American cities were shouting the ancient message that "the Jews have all the money" and that all bankers are Jews that "Jews control Wall Street" and that "Jews are greedy pigs."[58]

These messages of hate were transmitted by all of the electronic equipment available to the crowd and promptly given wider circulation by the media. This illustrates the "down" side of the use of cell phones and texting. The "flash mob," as sociologists call such social movements, communicate by means of cell phones and their derivatives.Texting can therefore be integrated into a campaign for or against anything and can be organized without much notice.

The use of cell phones also changes the manner in which reality is understood. For example, workers who have cell phones can always be reached by some boss at any time. In addition there are devices which can locate the user of a cell phone, both at present and in the past. This means that the line between private and work life has become blurred, particularly for those who are most dependent on their jobs.[59]

VII.

"Wiretapping" is a method by which a third party can overhear conversations of telephone users. When telephone exchanges were mechanical, technicians could link circuits together, routing the call to a third party. Today this can be done remotely by computer, which makes it impossible to know whether a line is being "tapped." Even mobile phones can be accessed. It is also possible to collect call numbers, times, and duration of calls, which are normally used for billing purposes by the telephone company. This means that all calls to a specific number can be discovered by sorting billing records. Conversations can also be recorded by a third party with a receiver as far away as ten kilometers, or 6.2 miles. Even a satellite can be used to receive transmissions anywhere on earth.

Mobile phones allow the collection of information about a telephone's location. The government of the United States routinely monitors citizens' phone calls with the cooperation of the telephone company. This gradually leads to a police state in which all activities of the citizens come under the scrutiny of the police and the innumerable bureaucracies operated by the federal, state, county, and local governments. In short, these devices guarantee the end of privacy and individual freedom in the United States, because those empowered to use these devices cannot resist introducing them into all private conversations.

Indeed, there has always been a good deal of controversy concerning "wiretaps" ever since they were first used by law enforcement. Therefore the U.S. Supreme Court decided in 1967 in Katz v. the United States that police must have a warrant signed by a judge to carry out such a search. One year later, in 1968, Congress passed a law allowing police to wiretap in cases of criminal investigations. Then, in 1978, as a result of the "Watergate" scandal, Congress passed the Foreign Intelligence Surveillance Act, creating a secret federal court which issues wiretap warrants. [60]

All of these activities confront the Fourth Amendment to the U.S. Constitution, which requires that Americans be "free from unreasonable search and seizures." This means that police are not permitted to search first and then use the evidence against a defendant. [61]

To defeat the fourth amendment, Congress passed the Communications Assistance to Law Enforcement Act in 1994. This law requires telecommunications companies to make it possible for government to "tap" into all subscribers to a telephone. The American Council of Education appealed to the D.C. Circuit Court in this matter. That court ruled on June 9, 2006 to dismiss that lawsuit, so that there has been a 3000 percent increase in interceptions of telephone and internet data, including email, since then. [62]

VIII.

One of the truly important benefits of the telephone has been the development of the emergency three digit telephone number 911. This development began in Great Britain in 1937, when the British government sought to provide citizens with an avenue to notify authorities quickly and efficiently of an emergency. [63]

In the United States, the 911 emergency system began in 1960, when the President's Commission on Law Enforcement urged AT&T to experiment with this system. It was therefore first employed in Haleyville, Alabama, where it was deemed successful. Subsequently, numerous other American towns adopted the system, so that almost all American wire lines accommodate the 911 emergency number. As a result, the American population is largely familiar with 911 and will use it whenever needed. Unfortunately, there have also been a number of instances when 911 was used unnecessarily. [64]

Until the increase in wireless telephones became substantial, the emergency operator, receiving a call at Public Safety Answering Points, would ask the caller a number of questions so as to determine the caller's location and the nature of the call. This meant that some time was needed to activate the help that was called for. Therefore, municipalities upgraded their systems to so that the 911 call automatically transmits the caller's telephone number and the location to the Public Safety Answering Points. This allows emergency personnel to gather needed information quickly even if the caller cannot convey his location.[65]

The need for the automatic transmission of the caller's location became dramatically illustrated in 1993, when eighteen-year-old Jennifer Koon was abducted from a shopping center in Rochester, New York. She was able to call 911 from her hijacked car but could not say where she was at the time of the call. Furthermore, the car was moving and there was no way of determining its location. Koon was shot to death by her abductor.[66]

Since a majority of Americans have in recent years acquired cell phones and discarded land line telephones, Congress passed the 911 Act to establish 911 as the official, universal emergency number for both wireline and wireless telephone service. The congressional act was called "The Wireless Communication and Public Service Act of 1999."

The Wireless Communication and Public Safety Act has four principle directives. The first is that 911 was to replace all other emergency numbers. This use of 911 as universaleliminated confusion by prohibiting alternative numbers. The act also charges the Federal Communications Commission with the task of creating a nationwide end to end network, allowing all states to administer a wireless 911 system. The act also equalizes the liability protection from transmission errors or technical failures provided by the states with that of land line users. Fourthly, the act protects the privacy of wireless users by requiring carriers to obtain written permission from a customer before disclosing any location information other than in emergencies.[67]

The reasons for the establishment of this "end to end" emergency service were six factual findings which Congress considered as the "cornerstones" promoting the passage of the 911 Act. These findings were, first, that the "end to end" system would save many lives and billions of dollars in health care costs because it would reduce the emergency response time. Congress also considered the "end to end" service as a means of reducing death and injuries resulting from motor vehicle accidents, which have led to a great deal of suffering on the part of motorists involved in such crashes. The act also encourages investments in emergency communications systems, which would allow further improvements in such systems. Congress also viewed the "end to end" system as a means of meeting the nation's public safety needs. This usefulness of the system was proved during the events of September 11, 2001.

In sum, the act seeks to allow any person located anywhere in the United States to receive assistance at any time.[68]

Summary

Among ancient people, ram's horns and fires were used to make long distance communication possible. It was the study of electricity which later led to the development of the telegraph and telephone. In the United States, this led to the invention of telecommunications, which has been attributed to the painter Samuel Morse.Morse was undoubtedly the final contributor to a long list of others who also participated in the invention of the telegraph.

Among other benefits, the telegraph made it possible to create Standard Time, which helped railroads a great deal. The United States entered the First World War in part because of the infamous "Zimmermann telegram."The telegraph was also of great help to government and to such chain news organizations as the Associated Press.

As telephones became more and more popular, the number of subscribers increased, so that in 2011 the United States had 327 million telephones.

In 1915 Alexander Bell made the first coast to coast telephone call. This development allowed newspapers and magazines to make nationwide surveys, leading to the disastrous Literary Digest survey of 1936.

The biggest change in telephone use came about with the invention of the cell phone. This great advance also has a dark side, in that the use of the cell phone and its additional variations has led to automobile accidents, resulting in injuries and deaths.

Both wiretapping and the 911 emergency network have had considerable influence on the life of Americans. Wiretapping, despite its use as a tool to prevent terrorism, endangers the right to privacy.

Notes

1. John B. Carlson, "Lodestone Compass: Chinese or Olmec Primacy?" *Science,* vol. 4205, no. 189, (September, 1975), p. 753.

2. Alexander J. Field, "The Magnetic Telegraph, Price and Quantity Data and the New Management of Capital," *The Journal of Economic History,* vol. 52, no.2, (June 1992), pp. 401-413.

3. Frederic C. Lane, "The Economic Meaning of the Invention of the Compass," *The American Historical Review,* vol. 68, no. 3, (April 1963), pp. 605-617.

4. No author, "The Telegraph in America," *The Illustrated Magazine of Art,* no.6 (1853), pp. 354-358.

5. JoAnne Yates and Robert J. Benjamin, "The Past and Present as a Window on the Future," In: Michael S. Scott Morrison, Editor, *The Corporation of the 1990's: Information Technology and Organizational Transformation,* (New York: Oxford University Press, 1991), pp. 61-91.

6. Ibid. p. 46-47.

7. Tom Standage, *The Victorian Internet: The Remarkable Story of the Telegraph.* (New York: Walker & Co.), pp. 4-5.

8. David Landes, *Revolution in Time,* (Cambridge, MA:The Harvard University Press, 1983), p. 240.

9. David McCullough, "Reversal of Fortune," *Smithsonian,* (Septmber 2011), pp. 80-88.

10. Robert L. Thompson, *Wiring a Continent: The History of the Telegraph Industry in the United States, 1832-1836.* (Princeton, NJ: The Princeton University Press, 1947), pp. 90-91.

11. John Porter Brown, *Turkish Evening Entertainment*, (New York: Adamant Corp., 2004).

12. Philip B. MacDonald, *The Saga of the Sea: The Story of Cyrus W. Field and the Laying of the First Atlantic Cable*, (New York: Wilson-Erickson, Inc., 1937).

13. Michael O'Malley, *Keeping Watch: A History of American Watch*, (Washington: Smithsonian Institution Press, 1990), p. 105.

14. Ibid., p. 100.

15. Tom Standage, *The Remarkable Story of Telegraphy and the Nineteenth Century On Line Pioneers*, (New York: Walker Publishing Co., 1999), pp. 161-162.

16. Mary Alexander and Marylin Childress, "The Zimmerman Telegram," *Social Education*, vol. 45, no. 4, (April 1981), p. 266.

17. David Paul Nickles, *Under the Wire: How the Telegraph Changes Diplomacy*, (Cambridge, MA: Harvard University Press, 2003), pp. 193-195.

18. Richard Schwarzlose, *Nation's Newsbrokers: Vol. 1: The Formative Years* (Evanston, IL: Northwestern University Press, 1989), pp. 366-370.

19. Alexander Field, "The Telegraphic Transmission of Financial Asset Orders," *Research and Economic History*, vol. 18, (1998), p. 146.

20. Anteresa Lubrano, *The Telegraph: How Technology Innovation Caused Social Change*, (New York: Garland Publishing Inc., 1977), p.124.

21. Melville E. Stone, "Fifty Years a Journalist," *Proceedings of the First National Newspaper Conference*, (Madison, WI: 1913), p.51.

22. Niall Ferguson, *The House of Rothschild*, (New York: Penguin Books, 1999), pp. 64-65.

23. Upton Sinclair, *The Brass Check: A Study of American Journalism*, (Urbana, IL: The University of Illinois Press, 2003), p. 41.

24. Will Irvin, "The United Press," *Harper's Weekly*, (March 28, 1914), pp. 10-12.

25. Robert V. Bruce, *Alexander Graham Bell and the Conquest of Solitude*, (Ithaca, NY: Cornell University Press, 1990), pp. 229-230.

26. Maury Klein, *The Life and Legend of Jay Gould*, (Baltimore, MD, Johns Hopkins University Press, 1986), pp. 277-282.

27. T. A. Wise, "Western Union by Grace of FCC and A.T. and T.," *Fortune*, (March 1979), pp. 114-119, 217-228.

28. Robert Flanagan, "How Western Union Went from Bad to Worse," *Business Week* (January 28, 1984), p. 110.

29. G. Stanley Shoup, "The Control of International Radio Communication, *Annals of the American Academy of Political and Social Sciences*, vol. 142, (March 1929), p. 103.

30. Seth Shulman, "*The Telephone Gambit: Chasing Alexander Graham Bell's Secret*, (New York: W.W. Norton, Inc., 1988) pp. 111-113 and 121-123.

31. Ulrich Stock, "Das Reis-Phone," *Die Zeit*, (October 15, 2011), p. 22.

32. A. C. Monahan, "Bell Rings Down a Century," *Science News Letter*, (March 1, 1947), pp. 138-139.

33. Shulman, *The Telephone Gambit*.

34. U.S. Bureau of the Census, *Statistical Abstract of the United States*, (Washington, DC: U.S. Government Printing Office, 1984), p.557.

35. U.S. Bureau of the Census, *Historical Statistics of the United States*, (Washington, DC: U.S. Government Printing Office, 1975) pp. 783-784.

36. No author, "The Telephone in Retail Business," *Printer's Ink*, vol. 61, (November 27, 1907), pp. 3-8.

37. Philip Dougherty, "New Pitch to Spur phone Use," *The New York Times*, Sec. D, (October 23, 1985), p. 19.

38. No author, "Western Electric Getting Farmers to Install phones," *Printers Ink*, vol. 76, (July 27, 1911), pp. 20-25.

39. No author, "Farmers Telephone at Low Cost," *Chicago Daily Tribune*, (September 20, 1903), p. 54.

40. No author, "Telephone," *Think Quest*, (New York: Educational Foundation, 2010).

41. James MacGregor Burns, *The Lion and the Fox*, (New York: Harcourt, Brace and World, 1956), p. 281.

42. George C. Edwards and Stephen J. Wayne, *Presidential Leadership*, (New York: St. Martin's Press, 1985), pp. 204-205.

43. John Goyder, *The Silent Minority: Non-Respondents on Sample Surveys*, (Boulder, CO: Westview Publishing, 1987), Chapter 6.

44. Steven Thomma, "Study: Polls that Leave Out Cell Phones Skew Results Toward GOP," McClatchy Newspapers, (October 18, 2010).

45. Don A. Dillman, ""Reducing Refusal Rates for Telephone Interviews," *Public Opinion Quarterly*, vol. 40, (1976), pp. 60-78.

46. Richard R. John, *Network Nation: Inventing American Telecommunications*, (Cambridge, MA: Harvard University Press, 2010), p. 407.

47. Anthony M. Townsend, "Life in the Real-time City: Mobile Telephones and Urban Metabolism," *Journal of Urban Technology*, vol. 7 (2000), pp. 85-104.

48. James Harkin, "Life Lines," *New Statesman*, vol. 17, (September 15, 2003), , p. 774.

49. Arriana Bassoli, et. al. "Underground Aesthetics: Rethinking Urban Computing," *Pervasive Computing*, vol. 6, (2007), pp. 39-45.

50. John Urry, "Mobility and Proximity," *Sociology*, vol. 36 (2002), pp. 255-274.

51. Matt Richter, "Drivers and Legislators Dismiss Cellphone Risks," *New York Times*, July 19, 2009), p. A1.

52. U.S. Department of Transportation, *"Statistics and Facts About Distracted Driving*, (http://distraction.gov).

53. Matt Richter, "In Study, Texting Lifts Crash Risk by Large Margin," *New York Times*, (July 28, 2009), p. A1.

54. David L. Strayer, "A Comparison of the Cell phone Driver and the Drunk Driver, *Human Factors*, vol. 48, (2006), pp. 381-388.

55. Marcel Adam, Timothy A. Keller and Jacquelin Cynkar, "A Decrease in Brain Activation Associated with Driving When Listening to Someone Speak, *Brain Research*, vol. 1205, (2008), pp. 70-76.

56. No Author, National Highway Safety Administration, U.S. Department of Transportation, "Using Wireless Communication Devices While Driving," (2003), pp. 3-4.

57. Andrew Keh, "Dugout Phones Last Bastion of the Landlines," *The New York Times*, (October 22, 2011), p. SP6.

58. Editorial Staff, "Ignoring the Lawlessness," *Augusta Chronicle*, (October 30, 2011), editorial page.

59. J. Sherry and T. Salvador, "Running and Grimacing: The Struggle for Balance in Mobile Work," In: Barry Brown, Editor, *Wireless World: Social and Interactional Aspects of the Mobile Age*, (London: Springer Verlag, 2006), pp.108-120.

60. Whitfield Diffie and Susan Landau, "Internet Eavesdropping: A Brave New World of Wiretapping," *Scientific American*, (September 2008), p. 10.

61. Constitution of the United States of America, Fourth Amendment.

62. American Council of Education v. FCC, 451 F. 3rd 226, 38 Communications Reg. (P&F) 859.

63. Bertram E. Maas, "911 Assistance Emergency Call Systems," vol 8, no.1 *George Mason University Law Review,* (1985), p. 103.

64. T. J. Yung and B. K. Yamaoka, *A National Directory of 911 Systems,* (Washington, DC: Bureau of Justice Statistics, 1980), p.1.

65. Federal Communications Commission, "Implementation of 911 Act," (Washington DC: 15 F.C.C. R17079, paragraph 1, 20 Com. Reg. 489, 2000).

66. Davis Schneider, "Dial L for Location," *American Scientist,* vol. 93, no. 6, (November-December, 2005), p. 504.

67. Maas, "911 Emergency Assistance Call System etc.," p. 103.

68. 145 Congressional Record, H. 9858, (October 12, 1999).

Chapter Seven

Radio and Television

I.

So many men have contributed to the invention of the radio that the endless dispute concerning priority of ideas and developments can never be resolved. Suffice it to understand that the radio, and subsequently television, was the direct outgrowth of the creation of the telegraph and wireless communication. Included in the list of contributors to the radio are Gian Romagnosi, Hans Ørsted and André Ampère. All of these scientists experimented with electromagnetic physical phenomena. It was the theory of electromagnetism as proposed by Michael Faraday, Joseph Henry, James Clerk Maxwell, and Wilhelm von Belzhold, which became the forerunner of the radio.[1]

It was David Edward Hughes (1831-1900) who discovered that sparks would generate a radio signal that could be detected by listening to a telephone receiver connected to a microphone. He was succeeded by a host of additional contributors, including Heinrich Hertz (1857-1894), Nikola Tesla (1856-1943), and finally Guglielmo Marconi (1874-1937), who succeeded in creating "electric wave telegraphy," resulting in the award to Marconi by the United States Patent Office of a patent for radio in 1901.[2]

The consequences of this development for the United States were indeed dramatic and revolutionary. In 1901 the United States was chiefly a country dominated by many rural communities. Communication between these communities within the three million square miles of the country was hardly possible. It was therefore the radio which became the most important catalyst in promoting American unity from coast to coast as never before. In addition, the radio became a source of instant news about faraway events in a manner not heretofore known. This became evident when on April 15, 1912 the luxury liner Titanic struck an iceberg in the Atlantic Ocean and sank, leading to the deaths of more than 1,500 of her passengers.

The Titanic was advertised as unsinkable. The ship was equipped with two 1.5 kilowatt spark-gap wireless telegraphs. This telegraphy was at that time the latest, most advanced system of communication, so that it was used for the first thirty years of radio transmission. The two radio operators on the damaged ship sent messages to nearby ships asking for help before the Titanic sank. The ships La Touraine and Mount Temple simultaneously heard the distress signal of the Titanic at 10:25 p.m. on April 14, 1912. In addition, radio communication made it possible that the ship Carpathia was at least able to rescue some of the Titanic survivors from the lifeboats they were using after the Titanic had already sunk. The Titanic disaster may also led to the growth of RCA, the Radio Corporation of America, because David Sarnoff, who was working for the "Marconi Wireless Telegraph Company," claimed he picked up transmissions from the Carpathia concerning the sinking of the Titanic and the names of the survivors. These events, said Sarnoff, led him to recognize the importance of radio, so that he made his career with RCA.[3]

In 1933, Edwin Howard Armstrong, a professor at Columbia University, invented Frequency Modulation or FM radio. This allowed for the transmission and reception of a wider range of frequencies and the elimination of "static," which made a disturbing noise during all AM radio broadcasts. FM eliminated all background noises and was able to access many radio broadcasters who could now send music as if the performance had been in the same room as the listener. Armstrong did not enjoy his invention for long. Radio Corporation of America disputed his patent and succeeded in the courts leaving Armstrong with nothing. He killed himself on January 31, 1954.[4]

It was Lee de Forest (1873-1961) who invented the Audion in 1906, which was an amplifier of weak radio signals. He is also credited with the birth of public radio when, in 1910, he succeeded in broadcasting the opera Tosca from the stage of the Metropolitan Opera in New York City, featuring the Italian singer Enrico Caruso. De Forest also broadcast the first radio advertisement. Finally, de Forest succeeded in improving the recording of sound on film, which led to "the talkies," with major consequences for the actors of that day.[5]

In addition to its usefulness in promoting political candidates and furnishing entertainment, the radio became a major instrument for the advertisement of manufactured goods and services. In the 1920's, more that one quarter of Americans were employed in farming, so that farm families became an important target for radio advertisers when radio broadcasts first began in 1921. During the decade of the '20's, four devices, i.e. the telephone, the automobile, electric light, and the radio, were introduced into rural homes. Of these, the radio became the most popular, so that by 1940 more farms owned radios than owned automobiles, electric light, or telephones.[6]

Rural life is isolated. Therefore, radio opened the world to farm families for the first time. Unlike rural newspapers which dealt only with local events, radio allowed farm families to learn all about current events and national issues. In addition, the radio created a single American identity and a single American consumer culture. This came about when, in 1930, six hundred and eighteen radio stations broadcast into about fourteen million American homes and three national broadcasting networks had been established.[7]

By 1930, about one half of all urban households had radios and around 27% of rural homes had radios. By 1940, radio ownership rose to 70 percent in rural areas and 84% in urban areas. In the 1930's, there were still a number of rural areas without electricity, so that radio reception was not possible. Furthermore, many radios used vacuum tubes, which required large amounts of electricity not available in many rural communities. Yet, radio had become a form of entertainment for many American homes and in addition furnished everyone with weather forecasts, which were particularly important to farmers.[8]

In 1930, "Motorola" installed the first car radios in a Ford Model A car for $540, which was a large sum at that time. Twenty-two years later, in 1952, the German radio manufacturer "Blaupunkt" installed the first FM radio in a car. In 1963, several manufacturers installed all-transistor radios in their cars, followed by car tape players, disks and other devices.[9]

The radio revolutionized politics in America, because it allowed elected officials and candidates for an office to speak to large audiences at once. The first American president to make use of the radio was Warren Harding (1865-1923), who gave a speech on the radio in 1922. In fact, two years earlier, in 1920, a Pittsburgh radio station was the first to announce Harding as the winner in that year's presidential election.[10]

Franklin D. Roosevelt (1882-1945) was the 32nd President of the United States. He assumed office in 1933, when one fourth of the American people were out of work and millions barely survived. In this situation, Roosevelt used the radio to speak to the American people in 31 "fireside chats." These talks were aimed at the "average" citizen, with a view of explaining numerous issues which troubled almost everyone. Roosevelt used only common language and thereby reached a large audience. In fact, many people felt that the president was speaking personally to them, so that his words had a unifying effect which alleviated the suffering of so many citizens somewhat. It is of course the verdict of history that the great depression ended only because the Second World War brought with it massive employment, both in the armed forces and in the defense industry. Nevertheless, Roosevelt took full credit for the eventual recovery, so that he was elected president four times.[11]

Since then, politicians of every political opinion have used the radio as a means of accessing their constituents. This has continued despite the widespread use of television, and particularly because FM or frequency modulation radio came into being when Armstrong invented this enhanced form of communication.

That radio differs in its effect from television became evident when Richard Nixon, the Republican candidate for president of the United States, debated John Kennedy, the Democrat, during the 1960 presidential campaign. Nixon was rated "the winner" by a two to one margin among those who heard the debate only on the radio, while Kennedy "won" over Nixon among those who saw the debate on television.[12]

Candidates for office are not the only politicians using the radio to send their message to anyone willing to listen. Charles Coughlin, a Roman Catholic priest, was undoubtedly the leading radio politician of the twentieth century. Coughlin developed a mass ministry in Royal Oak, Michigan, and broadcast nationally in favor of fascist causes, supporting the German government prior to the entrance of the United States into World War II. Coughlin preached hatred of Jews for a decade, before his support for Germany and its Nazi policies became unpopular in light of American involvement in the war against the Axis powers.[13]

Since the days of Coughlin, numerous other "talk show" hosts have harangued the radio listener with their various messages. Some of these are political, others religious or concerned with one issue repeated again and again in an effort to persuade the listeners and others. In the 1920's, William Jennings Bryan was a famous orator on economic issues who used the radio to gather his followers and influence politics. Later, Joseph McCarthy, senator from Wisconsin, used the radio to whip up anti-communist hysteria, while Huey Long in Louisiana and George Wallace in Alabama proclaimed their anti-government obsessions with their populist messages. The effort to bring about desired political results has continued on American radio, as witnessed by the huge success of Rush Limbaugh, who broadcasts every day from 600 radio stations around this country.[14]

Since the earliest days of radio broadcasting, the radio has been a source of entertainment. On October 30, 1938, the actor Orson Wells broadcast a play adopted from H.G. Wells' story "War of the Worlds."The broadcast simulated a series of news bulletins to the effect that "aliens" had landed in New Jersey and were advancing on New York. Some listeners panicked, believing this charade. Many called police, only to learn that this was entertainment. While earlier reports claimed that the Wells broadcast caused mass hysteria, later analysis indicated that only a few "fell" for this deceit which made Orson Wells famous.[15]

Most popular radio shows before television were comedies, of which "The Jack Benny Show" ran the longest, beginning in May 1932 and continuing on television until 1965, because Benny wanted to retire and not because the ratings had declined. Benny was born Benjamin Kubelsky, and was married for many years to Sadie Marx, who was billed as "Mary Livingstone" in his weekly program consisting of jokes, innuendoes, songs, sarcasms, and a variety of hilarious entertainment supported by a talented cast.[16]

"Fibber McGee and Molly" was also a popular radio program starring a husband and wife team, James Jordan and Marian Driscoll. This show began in 1927 on a Chicago radio station and continued until 1961, when Driscoll died of cancer. That show included standing references to McGee's closet, a neighbor called Gildersleeve, and a host of other hilarious characters which became part of popular culture for 35 years.[17]

Likewise, "The Goldbergs" became one of the most popular radio shows, beginning in 1929 and continuing into television until 1956. This show depicted a poor Jewish family "bossed" by a stereotypical "Jewish mother," the actress Gertrude Berg, who was cast as "Molly Goldberg." A domestic comedy, the show dealt with non-controversial material depicting everyday life. The popularity of the show stemmed mainly from the ability of Americans of any ethnic group to identify with the events happening to the fictitious Goldberg family.[18]

Radio entertained millions of Americans with many additional shows such as Amos 'n Andy, Lum and Abner, Easy Aces, Vic and Sadie, and more. Some ended with the advent of television; others survived and became television entertainment as well.

Among these was the weekly broadcast of the Metropolitan Opera, sponsored every Saturday by Texaco and still on the air in 2012. These broadcasts were sponsored by Texaco for 63 years, starting in 1940 and ending in 2003. The broadcasts continue today with different sponsors.

The most recent development in radio broadcasting is satellite radio, which was first introduced in 1990 in Washington, DC. This radio service broadcasts with satellites to listeners with special receivers who pay subscription fees. In 2000, "Sirius" launched one satellite and later two more. "Sirius" also succeeded in signing Howard Stern, a controversial and provocative talk show host. In addition, Sirius has arranged to broadcast NFL games and other sports.[19]

Although television is found in almost all American homes, radio continues to be found everywhere, in homes, cars, workplaces, and in handheld devices.

II.

On July 3, 1957, Philo Farnsworth appeared on a television program. It was the only appearance of the inventor of television on the medium he had created. Farnsworth was a mystery guest on a TV show called "I've Got a Secret."A panel of "celebrities" was told his name but was unable to guess at his role as the inventor of electronic television. In short, he "stumped" them, as none of the panelists had ever heard his name before.[20]

To this day, in 2012, few Americans have ever heard that name, although 98% of all American homes have a television set and are using the Farnsworth invention. This came about because Farnsworth was a naïve young man when he succeeded in producing the first television in a loft in San Francisco on September 7, 1927. By 1929, he transmitted the first televised image, a straight line.

This achievement was the product of a genius mind, for Philo Farnsworth was only fourteen when he sketched on several blackboards in his school how an electronic television might be constructed. Unfortunately for him, Farnsworth was visited by a Russian born physicist who was working for the Radio Corporation of America in an effort to develop television. Vladimir Zworykin saw what Farnsworth had accomplished and used that knowledge to produce a television for RCA. That company stole Farnsworth's ideas and tried to steal his patents but after lengthy litigation the patent office sided with Farnsworth in

1934. Nevertheless, Farnsworth's name was not included as the inventor of television in any American encyclopedia. Instead, manipulations by RCA led encyclopedia editors to believe that Zworykin was the inventor. RCA did pay Farnsworth $1 million to settle the dispute about Farnsworth's patents. Farnsworth then established a television production company, but he was never able to match RCA in producing and distributing his invention.[21]

That invention led to the creation of an entire television world by television producers. This world is an abstract of reality and therefore portrays people and events in a manner which is largely fictitious. No doubt, a considerable number of viewers believe they can discern the difference between television fiction and reality. Nevertheless, the dictum that "If men define situations as real they are real in their consequences" applies. Undoubtedly there are some who are not always certain concerning the state of reality, a condition which has led to considerable dispute among philosophers and which Sigmund Freud attributed to all children.[22]

That television creates a secondary world can easily be demonstrated. For example, the portrayal of women and men in advertisements but also in movies and television "shows" leaves viewers with the notion that the women and men one meets in real life ought to be treated as portrayed on television. Accordingly, sexual attraction is greatly emphasized in almost all television programs, although few Americans look or talk as these programs indicate.

The manner in which gender is treated in television episodes and in advertisements appearing on television is a good example of the gap between reality and the television world, although much of what is depicted on television influences real life situations. Because the average American household watches more than seven hours of television each day, the influence of TV on American life can hardly be exaggerated. This includes gender stereotypes. Therefore, the constant depiction of women dealing with dirt as seen in TV advertisements can hardly contribute to enhancing female status in this country.[23]

If television viewers can be convinced that a product sold by a TV housewife has value, then it is likely that the stereotypes of women cleaning toilets or shoving around dirt will create the impression that women are what the TV advertisers portray. This may be labeled "the happy housewife" stereotype.

Although agitation for full female participation in the television industry has continued since the 1960's, a 2011 report by the Center for the Study of Women in Television and Film at San Diego State University reported that female characters in movies made up only 33% of actors appearing in films released in 2011. In January of 2011, the Center found that women made up only 18% of "behind the scenes" directors, producers, writers, cinematographers, and editors.

Nevertheless, television has given women opportunities which prior to 1984 did not exist. In that year, Oprah Winfrey became the first woman to broadcast the news on WLS-TV in Chicago, Illinois. She went on to become the richest black woman in America, with an estimated worth of $2.7 billion. Winfrey is of course an exception, but the presence of women in television broadcasting is by no means unusual in 2012. American University's "Women and Politics Institute" and other students of gender in broadcasting found that in 1987 men reported 73% of news stories on television. In 2007 this had changed, so that beginning then men alone reported 48% of news stories, women alone reported 40% of news stories, and one woman and one man together reported the remaining 12 percent of the time. Evidently, television has been a true vehicle for the achievement of gender equality in America.[24]

This drive for gender equality in the television broadcasting industry led to the appointment of "Katie" Couric to be news anchorwoman on CBS in 2006. She continued in that position until 2011, when she was discontinued because her ratings fell. Nevertheless, she became a talk show host at ABC in September of 2012. Barbara Walters, Soledad O'Brien, Diane Sawyer, and a host of other women have assumed so many leading positions on television that male prerogatives in that medium have definitely come to an end.[25]

The evidence is that television has contributed a great deal to women's equality, even as advertisers and others continue to repeat some of the old stereotypes concerning women, which persist in an age when women have achieved professional and political attainments utterly unknown only one generation ago.

It is commonly believed that only members of "the fairer sex" are the victims of sexism. Yet, an inspection of the manner in which men are portrayed on television and in other media reveals that entrenched views concerning the status-role of men continue. Men are constantly berated by advertisers to enhance their sexual prowess by buying all kinds of pills and other products reputedly overcoming "ED" or erectile dysfunction. This kind of advertising supports the adolescent view that masculinity is invariably tied to "performance" and that those who by reason of age, physical condition, or interest do not "perform" adequately are somehow not men. Concomitant with that propaganda is the ludicrous complaint that "men are interested in only one thing."

While women have largely succeeded in defeating stereotypes concerning female incapacity and other prejudices, men have not succeeded in overcoming these antiquated notions. There are no male counterparts to the National Organization for Women and other women's defensive groups. Therefore, men continue to be portrayed on television and particularly in advertisements as interested only in sex, beer, sports, and their lunch buckets. The facts are that innumerable men have innumerable interests from anthropology to zoology and so much in between.

Old age is also distorted on television. Although the old are the fastest growing age group in America, they seem to be almost nonexistent according to television. In 2012, over 56 million citizens received old age social security money, constituting about 15% of the American population. Yet, old people are only 5.6% of televised people.[26]

Among the few programs concerning the lives of old Americans was "Golden Girls," which ridiculed the lives of several senior women. Likewise,

"Grumpy Old Men" was shown on television repeatedly. This movie recites all the stereotypes and prejudices concerning old age that are commonly believed. According to this film, old men are nasty, angry, and belligerent, and their lives are boring and lonely. Because old age is hardly shown on television and then only in a negative light, the medium supports all the unreal beliefs concerning this stage of life.

It is an axiom of American social life that one refrain from discussing politics and/or religion if interested in maintaining friendships or conversing in polite society. However, this maxim does not apply to television broadcasting. On the contrary, both politics and religion are constant features of television, and both receive an inordinate amount of time in this medium.

In a presidential election year, such as 2008 and 2012, politics dominate the news presented by the several television networks. It is widely believed by American television audiences that these networks are biased in favor of either the "conservative" or Republican ideology or in favor of the "liberal" or Democrat ideology. Therefore, viewers who oppose one or the other political point of view will attribute bias on the part of the network in favor of those with whom they do not agree when viewing the news. Yet, a detailed study by Joel Turner discovered that selective perception, rather than bias on the part of newscasters, distorts the news broadcasts delivered by CNN and Fox News. CNN is widely viewed as "liberal" and Fox is widely believed to be "conservative." Turner found that ideological bias in the news is hardly discernible, but that the label of the news organization leads viewers to attribute political bias to the story even when the broadcasts of both networks are substantially the same. This means that the same story, told the same way, is believed to be skewed in favor of "conservatives" or "liberals" depending on the views of the consumer and not the producer. Turner found that "individuals can watch virtually identical stories on CNN and FNC yet still derive different ideological signals from each network."[27]

Politics are closely tied to the perception voters have of candidates for political office. Those perceptions entered the living rooms of American voters with the advent of television. Therefore it is not surprising that the incumbent usually has an advantage over the challenger, because television gives him greater access to the voters than can be true of the challenger, who does not hold the targeted office.

A study by Markus Prior demonstrates that in U.S. House elections, incumbents have had a distinct advantage ever since the rise of television as an almost universal component of the American home. Prior writes that "...the growth of television increased the incumbency advantage in the U.S. House elections in the 1960's."[28]

There is a good deal of evidence that voters will expect their representative to be helpful in the future based on the frequency with which their name is recognized in the media, and particularly how often the office holder is seen on television.[29]

III.

American children watch TV to such an extent that a considerable amount of TV advertising is directed at them. It has been estimated that children aged four to twelve influence over $165 billion in spending, and six to twelve year olds directly buy $24.4 billion worth of merchandise each year. This therefore has created a great deal of children's programming and a good deal of competition among networks, leading to airing of children's programming seven days a week.[30]

It has been shown that beginning at age four, children have an understanding that others may have wishes differing from their own. This is a prerequisite to children's understanding of advertising. Therefore advertisers are also dependent on the amount of time children spend watching TV. Depending on their age, children watch television at least seven hours a week at age two and as much as 14 hours a week at age five. Children age eight to eighteen watch television 53 hours a week, or 7.5 hours a day. Researchers have also found that even some nine month old infants regularly watch television.[31]

Studies have shown that violence and sexual content are included even in programs aimed at young people. This means that children can be influenced in their daily lives by such material, which is associated with antisocial and unhealthy behavior.[32]

According to several reviews of television programs viewed by children, it appears that by the age of eighteen the average American will have viewed 16,000 murders and 200,000 other violent acts. Since viewing times have increased recently and more violence is shown on TV than ever before, it is fairly certain that these statistics will soon be revised upwards.

Children are more likely to watch cartoons than other kinds of TV entertainment. This leads the uninitiated to believe that children therefore see little violence. Yet, observation of cartoons reveals that cartoons include a great deal of violence, which is presented as trivial and hilarious. This teaches children that violence is not serious, a message also promoted by numerous war movies and "western" episodes in which victims of shootings seem to ignore serious wounds or "brush them off" with heroic gestures.

All of this has lead the American Psychological Association to conclude that children who have seen this mayhem will become less sensitive to the pain and suffering of others, be more fearful of the world around them, behave in aggressive and harmful ways towards others, and be less likely to see anything wrong with violence.[33]

There are those who believe that violent and sexual content that children view on television or elsewhere has a major impact on children's attitudes and conduct. Of these two problematic kinds of programs, violence is more common than sexual exhibitions. Although the relationship between on-screen violence and real violent conduct has not been proved, it is agreed among researchers that pictured violence does have emotional consequences for children and adults. There is also some evidence that exposure to sexual scenes on television can lead to teens' initial sexual behavior, which is usually casual and risks both disease and unwanted pregnancies.

By the same token, children who watch family oriented TV programs are led to believe that families in real life show concern and support for each other. Therefore some parents have installed blocking devices on their television preventing children from watching sex and/or violence.

IV.

Television has become most influential in causing viewers to spend a great deal of money attempting to improve their attractiveness. This can be verified by considering the amount of money spent on cosmetics, plastic surgery, and clothes. While both men and women are affected by the discrepancy between television portrayals of good looking people and the reality of their own appearance, men do not increase their spending appreciably to enhance their body image. Women, however, are far more affected by the real-ideal self discrepancy and are therefore much more willing to adjust their consumption to their need to improve their appearance.

It is evident that viewing television over a long period of time has a good deal of effect on understanding social reality. This means that many television viewers believe that what they see on television represents reality. Since television promotes a distorted and fictitious view of the world, many television viewers are led to believe that they live in a world which does not really exist. For example, television grossly exaggerates the number of lawyers, doctors, and police in the United States. Actors are also constantly shown, although the proportion of such occupations to all occupations is relatively small. Likewise, tele-vision distorts consumption of alcohol, drugs, and sports as well as the number of adults who have had plastic surgery or are using beauty enhancing products.[34]

Television programs invariably display attractive actors and actresses, including those who recite the news of the day or conduct interviews or participate in talk shows of all kinds. This leads heavy viewers of television to believe that they are watching the average American body and face. Women, even more than men, will conclude that being thin is important and that they need to alter their appearance by means of plastic surgery. The American Society for Aesthetic Plastic Surgery reported that in 2011 women had 8.6 million procedures, which constituted 92% of all procedures. Included in these operations were over 318 thousand breast augmentations, 289 thousand liposuctions, 152 thousand eyelid surgeries, 145 thousand abdominoplasties, and 138 thousand breast reductions. This means that there was an increase of 164% in plastic surgery among women since 1997. Although men had only 8% of the total number of plastic procedures, the number of such procedures for men increased by 88% since 1997. Americans spent $10.7 billion on cosmetic procedures in 2010, of which $6.5 billion was spent on surgery.[35]

The Greek word "cosmetic" is derived from "cosmos," meaning order. According to a March 2012 research report on the cosmetics industry in the United States, American women spent $13 billion on beauty and fragrance products in 2011. That industry employed 175,427 people working in 89,378 businesses that year. These products are sold in department stores, drug stores, and specialty beauty stores. It has been estimated that the beauty industry will grow at an annual rate of 7.8% in the ten years ending in 2017. During that decade, the gross domestic product of all goods and services produced in the United States will grow about 1.95%.[36]

Included in these expenditures are lipsticks, blush, eye shadow, and foundation, sold by such companies as Maybelline and Clinique. Fragrances of all kinds are also sold with such French sounding names as "Chanel #5" and "L'Oreal." These companies also sell deodorants, sunscreens, and depilatories or hair removers (from Latin meaning "capped").

Men are also affected by television and other media efforts to define the "ideal" American man. A popular singer, Joe Diffie, sang a song called "Pickup Man," which refers to a man who meets numerous women who seek his company because he drives a "pick up truck."

Ford trucks are advertised on television as symbols of masculinity. Here a deep voiced man, Dennis Leary, refers to dirt and hard work on the part of Ford truck owners as a truck is seen smashing the ground with violence and implying that that the truck driver is as "tough" as the truck itself.The commercial seeks to link bravery and adventure to truck ownership and includes the phrase "the eagle has landed," which was used by Neil Armstrong on landing on the moon in 1969. This commercial is directed at the "working class" with the comment, "you are not making money pushing a pencil or hand modeling. You are actually working for every dollar and need your truck." Here it is implied that white collar men are not really masculine. No doubt there are men who buy a truck because they are persuaded by television that driving a truck makes them "rugged," aggressive, strong "he" men in the American tradition.[37]

Commercial television sells products which seek to emphasize male identity. Because the average American television viewer sees 31 hours of television a week, including about five hours of commercials, these commercials define gender norms. Men in these advertisements are heterosexual, competitive, aggressive individualists who are emulated by other men. The television male is strong and successful and derives his reputation from his workplace. These men dominate women and are admired by them. This scenario is of course created by American culture and is not biological. Masculinity is therefore defined, not by what a man is, but what is he is not, that is, he is not like a woman.[38]

Men in these advertisements are usually seen in public and not at home. Women are depicted in domestic scenes but men are seen "on the job," where they are competent, dominant, and experienced. Advertisers also like to use male voices in selling their products, although in the 21st century, these views of men and women are outdated, as women have achieved considerable autonomy and competence in many occupations as well as in leadership roles.[39]

Television commercials also project male roles as carefree, unencumbered by women, families, or any obligations, as they drive the advertised car or drink beer in an outdoor setting while taking physical risks which the car and the beer help them mightily to overcome. These pseudo men master technology, nature itself, and all challenges, while gaining access to desirable women and the admiration of important men. The message is that alcohol, mechanical tools, and life insurance make the man. In short, it is consumption of these products that confirms masculinity.[40]

Beginning in the 1990's, more and more men have remained single or are divorced, and therefore are forced to buy food and other products sold in grocery stores which were at one time the nearly exclusive realm of female customers. Single men are also more open to buying all kinds hair restorers, fragrances, and "after shave" lotions which reputedly attract women. The advertisement seeks to convince the male user that the substance he is using will make an ordinary man a magnet for the opposite sex. It is the product, rather than the man which attracts. Therefore, the erstwhile gap between male and female grooming practices is decreasing as sales of male skin care products increase.[41]

V.

Every television news broadcast includes a segment on sports. Sports broadcasts include brief segments of games played earlier that day together with an analysis by a sports announcer. According to the Bureau of Labor Statistics, sports announcers earned an average annual salary of $27,000 in 2012.[42]

Sports are included in every news broadcast because sports are one of the most important forms of television entertainment. This is particularly true of football, which gains a larger audience than any other television program. On February 5, 2012, more than 111.3 million Americans watched the "Super Bowl" football game between the New York Giants and the New England Patriots. No other television broadcast reaches so large an audience.[43]

There can be little doubt that the National Football League owes its existence and its income to television. Prior to the advent of television, football was almost entirely a college game, with little interest shown the underpaid, almost unknown, professional football players. Although the first football game was televised in 1939, it wasn't until 1945 that television became a form of mass communication, providing sports of all kinds, and particularly football, with a mass audience.[44]

Basketball is at least as popular in the United States as football, although basketball tournaments are played by college students or professionals posing as college students. In1939, the National Collegiate Athletic Association was founded so as to promote college basketball as a national sport. That was before the advent of commercial television. Since then, television has done for basketball what it has also done for football.Since 1969, television has broadcast the national championship tournament, which is labeled "March Madness," a term coined by Ohio State basketball coach Harold Olsenin 1938.[45]

Whether football, basketball, or the Olympics, the viewers appear to satisfy several needs through viewing sports. One of these is a sense of belonging. There are football fans who dress in the colors of the team they support. Some behave in a violent and bizarre manner when viewing or attending a football game. In some cases, football fans have become violent in their anxiety to relieve tension and aggression. For example, in 2001, Cleveland, Ohio football fans threw numerous bottles and other objects onto the field after officials overturned a call, leading to a Cleveland loss. Some fans were so enraged by that loss that they tried to assault the Jacksonville team.[46]

In all American communities there are "sports bars," which are so called because they feature several TV sets allowing customers to watch different football and other games while drinking. These bars pay the NFL $1,000 to $2,600 for the Sunday Ticket Program, which allows a bar to show every NFL game during a five month season. The Sunday Ticket program is also available in private homes for $179. Subscribers can see 14 games each week.[47]

The Olympic Games are televised worldwide and draw an immense audience. The 2008 Olympic Games in Beijing, China were watched by 217 million American viewers over a period of 3,500 hours of coverage. The Olympic Games in London was televised around the world. Forty different sporting events were broadcast live with a view of programming 5,000 hours of television coverage.

Because television is so important a part of many Americans' lives, an entire popular literature accompanies television programs from news to sports and so-called "soap operas." All Soap Scoops advertises weekly opinions, rumors, scoops, and links for soap opera fans. Cable World is a professional magazine which focuses on operators, vendors, and programmers. Entertainment Weekly discusses TV shows, publishes interviews with "celebrities" who are usually actors, reviews TV movies, and lists music and books of interest to TV watchers. N Zone Magazine publishes reviews of TV shows and interviews with actors or others of importance to television. Reality TV shows are reviewed by Reality TV Magazine, with commentary on a number of reality TV shows such as "Dr. Phil," "Survivor," "The Apprentice," "The Osbournes," and "TV Contest." This magazine publishes interviews with contestants on such shows as "Wheel of Fortune." Real Screen includes information about business opportunities in television and publishes articles about lifestyle programming.Smart TV contains a buyer's guide covering devices that enhance smart homes. Soap Opera Digest discusses day time and prime time soap operas. Also included are interviews with TV "personalities" as well as on camera "happenings" which are of course not real but are taken as such by some viewers. Television Weekly covers all aspects of the TV business including programming, production, cable and satellite, finance, and technology. TV Guide shows local listings, the latest gossip about the latest shows, movie reviews, and exhibitions of celebrities.

This list is by no means inclusive. There are numerous additional periodicals dealing with all aspects of television. Furthermore, all daily newspapers have a television page in which they publish the daily television schedule.

Because television and radio broadcasters depend on advertisements to remain in business, they generally use Nielsen Media Research to discover whether or not a program should remain "on the air" or be cancelled. Nielsen "ratings" are an invention of Arthur Nielsen, who founded this company in the 1920's and expanded it into radio research in the 1930's. In 1950 Nielsen moved to television, using a number of statistical devices to measure the size of audiences for any program broadcast at any time. Since advertising rates are influenced by these ratings, Nielsen provides statistics on specific demographics such as the age, gender, or wealth of a potential audience viewing the advertisement of a product.[48]

VI.

Not all television broadcasts are commercial. There is also educational television, which developed slowly beginning in the 1950s. At that time television audiences had reached millions, as visible by the decline in movie attendance. For example, in 1946 the American film business grossed $1.7 billion. By 1962 the film industry declined to one half of its previous income, grossing only $900,000,000.[49]

Educational television differs from commercial television in several ways. First, its programs are arranged in series leading to cumulative learning. The programs are planned in consultation with educational advisors. They are usually accompanied by textbooks or study guides and the broadcasts are evaluated by teachers and students.[50]

Because educational television is "live," it cannot be previewed nor be interrupted for questions. In that regard it differs from movies, particularly as it is broadcast only once and must be seen at that time. Educational television also allows students to experience foreign cultures and historical events as portrayed on one or more history channels. Educational television has also had the advantage of letting teachers ask more questions than is common in non-televised situations. Some claim that students develop more thinking skills when taught with the help of television programs.[51]

There are some educational programs on commercial television. The first of these was the "Johns Hopkins Science Review," followed by "Walter Cronkite's Universe." These programs did not last long, because only about ten million ever viewed them, a number too small to be sustained by profit oriented commercial stations. This failure led to a public hearing in Congress concerning the future of educational television. As a result of these hearings, the Federal

Communications Commission reserved 162 Ultra High Frequency and eighty Very High Frequency channels for educational television in 1952.[52]

The first educational television station was in Ames, Iowa. It was licensed to Iowa State University in 1950. It continued until 1994, when the trustees sold the station. It was also in Iowa that TV Schooltime was broadcast from 1952 to 1970, followed by Iowa Television Science, which lasted only one year. Then a conglomerate of Midwestern educational television broadcasters developed "Adventures in Science" and "Exploring the World of Science" until 1968. These programs and others resulted in the finding that students in the television viewing classrooms averaged higher test results than those in the non television setting.[53]

In the 1970's, local efforts at educational broadcasting were gradually absorbed by the Public Broadcasting System. This system broadcast NOVA, which has become one of the longest running shows on television. It is produced by WGBH in Boston and deals with the creative process in science.

Seeking to overcome science illiteracy among so many American students, PBS developed Cosmos, which emphasized astronomy and presented some of the work of Carl Sagan. Biology and a variety of cultures were also taught in that program. Einstein's Universe was another program developed by PBS. The narrator was the actor Peter Ustinov, rather than a scientist. The idea was to emphasize the human element in the scientific endeavor, which showed that science is an activity in which anyone can participate and which anyone can enjoy.[54]

In the early 1960's, a number of school districts adopted closed circuit television, including intra-school networks, which in some school districts meant that the entire curriculum was related to a closed circuit television network. As a result, 94% of teachers surveyed preferred teaching science with television.

The Discovery Channel, a cable network, also offers a number of programs for classroom use. The channel offered teachers a set of support materials to use in conjunction with the televised material, including lesson plans and references to online resources. Neither the Discovery Channel nor any television program replaced teachers, as some early enthusiasts believed. However, television has become an invaluable aid to teachers, beginning in elementary school and continuing into higher education. Television has developed interest in subjects many students never considered until television made it more understandable. This is particularly true of science programs. Television also promoted interest in history and a number of social issues which entertainment television could never realize.

VII.

Employment in the television industry is not large nor lucrative. There are about 62,000 announcers working in American television. Their average annual pay is $27,000. Announcers present music, news, and sports, and interview guests.

They also may present commentary on important local events. Because they are seen on TV every day, they are often invited to be masters of ceremonies at public events, but also at weddings, parties, and clubs.

Radio and television announcers are usually college graduates who gained experience by announcing in the college radio and TV stations. The Bureau of Labor Statistics has estimated that announcer job openings will grow by 7% by 2020.

Cameramen are essential in the TV business. Television employs 17,500 cameramen, who earn an average salary of $44,130. The range of income is fairly long, in that beginners earn only $21,000 while senior cameramen can earn $83,000.

Producers and directors are also employed in the television industry. The 122,500 jobs in those categories pay a median income of $69,000 annually. Producers and directors create television shows by interpreting a writer's script. These are high stress jobs because the work arrangements are usually short, ranging from one day to a few months. This means that many directors and producers hold a second job to make a living. The Bureau of Labor Statistics has estimated that employees in both categories therefore work for television on an assignment basis. Their median income is $54,000 annually, although some exceptional writers who produce whole series can earn millions. About 146,000 writers work full time in this occupation. There is a great deal of competition for full time salaried writing jobs.

TV executives may be executive producers who earn an average salary of $72,000 or they may be account executives. Executive producers are generally recruited from experienced directors, writers and/or editors. Ordinarily executive producers have a master's degree in television production broadcasting or in communications. The most important skills demanded by executive producers are networking with others in the industry and the ability to market creative concepts.[55]

Account executives earn an annual average of $42,000. Most account executives earn their income through a combination of salary, commissions, and bonuses. Their job is to constantly obtain more advertising and sponsorships while maintaining existing accounts.There are some account executives who earn much more than the average because they wok for a national television network usually located in New York or Los Angeles.

Summary

Numerous scientists contributed to the development of the radio, although Marconi was granted the first U.S. patent for this invention in 1901. The radio made possible instant news from coast to coast and was instrumental in broadcasting the Titanic disaster in 1912. This led to David Sarnoff's career at the Radio Corporation of America. FM radio was invented by Armstrong in 1933 and De Forest improved the talking movies and the vacuum tube. Radio became a means to relieve the isolation of rural families as well as a vehicle for political activities.

Television was largely invented by Philo Farnsworth as well as other contributors. This has led to a television world which deals chiefly in unreality as a source of entertainment. Television is dependent on advertisement for its income, so that a good part of television is directed at consumers, including children.

Television broadcasts stereotypes regarding gender, age, occupations, and ethnicity. It is also a major catalyst for the support of sports on the part of a large number of Americans. In addition, television is a vehicle for the dissemination of political propaganda as well as a source of entertainments of all kinds. This includes so-called "soap operas" which are exceedingly popular. In addition to commercial television, there is also a good deal of educational television available.Television "shows" are rated by means of statistical investigations concerning the size of each audience of any particular broadcast.

Because of the competition to be employed in television, the salaries paid are generally poor, with the exception of a few "superstars" who earn millions.

Notes

1. Ivor Hughes and David Ellis Evans, *Before We Went Wireless*, (Burlington, VT: Images from the Past, 2011).

2. Lucien Poincare, *The New Physics and Its Evolution,* (London: Paul, Trench & Co., 1907), Chapter 7.

3. Alexander B. Magoun, "Pushing Technology: David Sarnoff and Wireless Communication," *Conference on the History of Communications*, (St. John, Newfoundland: July 26, 2001).

4. Max Millard, "Lee DeForest: Class of 1893, Father of the Electronics Age," (Northfield: *Mount Hermon Alumni Magazine*, October 1993).

5. Charles Susskind, "Armstrong, Edward Howard," *Dictionary of Scientific Biography*, (New York: Charles Scribner's Sons, 1970), pp. 287-288.

6. Ronald R. Kline, *Consumers in the Country: Technology and Social Change in Rural America*, (Baltimore: Johns Hopkins University Press, 2000), p. 2.

7. Steven Craig, "How America Adopted Radio," *Journal of Broadcasting and Electronic Media*, (June 2004), pp. 179-195.

8. Milton Eisenhower, "Uncle Sam Chats with his Dairymen," *Radio Age: The Magazine of the Hour*, (October 1926), pp. 37-40.

9. Justin Berkowitz, "A History of Car Radios," *Car and Driver*, (October 2010), p. 1.

10. No author, "This Day in History," http://www.history.com/this-day-in-history.

11. Diana Mankowski and Raissa Jose, "Flashback: The Seventieth Anniversary of FDR's Fireside Chats," (Chicago: The Museum of Broadcast Communications, no date), p.1.

12. Edward W. Chester, *Television and American Politics*, (New York: Sheed and Ward, 1969), pp. 59 and 120.

13. Donald I. Warren, "Radio Priest: The Father of Hate Radio," *Michigan Sociological Review*, vol. 11, (Fall 1997), p. 114.

14. Ibid., p. 114.

15. Glen Levy, ""War of the Worlds: Top Ten Hoaxes," *Time,* (March 16, 2010), n.p.

16. John Dunning, *The Encyclopedia of Old Time Radio,* (New York: Oxford University Press, 1998).

17. Jim Jordan, "Radio's Fibber McGee" *The New York Times,* (April 8, 1961), p. 10.

18. Pau Lomatire, "Have I Got News for You About Molly," *The Palm Beach Post,* (June 18, 1994), p. 1D.

19. Ellen Sheng, "Sirius Bolsters Lineup to Battle FM," *Wall Street Journal,* (June 2, 2004), p. 1.

20. Paul Schatzkin, *The Boy Who Invented Television,* (Silver Springs, MD: Tamcom Books, 2002), pp. 20-21.

21. George Everson, *The Story of Television: The Life of Philo T. Farnsworth,* (New York: W.W. Norton & Co., 1949).

22. William I. Thomas, *The Child in America,* (New York: Alfred Knopf, 1929), p. 22.

23. John Bardwick and S. Schuman, "Portrait of American Men and Women in TV Commercials," *Psychology,* vol. 4, no. 4, (1967), pp. 18-23.

24. Kathleen Ryan and Joy Chavez, "Media Report to Women," *Electronic News,* vol. 4, no. 2, (2010), pp. 97-117.

25. Lauren Sher, "Katie Couric Returns to ABC," *Internet Ventures,* (June 6. 2011), p. 1.

26. U. S. Bureau of the Census, Summary File, Table P12.

27. Joel Turner, "The Messenger Overwhelming the Message," *Political Behavior,* vol. 29, no. 4, (December 2007), p. 455.

28. Markus Prior, "The Incumbent in the Living Room," *The Journal of Politics,* vol. 69, no. 3 (August 2006), p. 668. "Incumbent" is derived from the Latin "to lean on something."

29. Bruce E. Cain et. al. "The Constituency Service Basis of the Personal Vote for the U.S. Representatives and British Members of Parliament," *American Political Science Review,* vol. 78, no. 1, (1984), pp. 110-25.

30. Charles Stern, "PBS Tries to Keep Egg in Nest," *Variety,* (June 1998), pp. 21-24.

31. Frederick Zimmerman, Dimitri Christakis and A.N. Meltzoff, "Television and DVD/Video Viewing in Children Younger than Two Years," *Archives of Pediatric and Adolescent Medicine,* vol. 161, no.5, (2007), pp. 473-479.

32. J. D. Brown, et. al. "Sexy Media Matter: Exposure to Sexual Content in Music, Movies, Television and Magazines," *Pediatrics,* vol. 117, no.4, (2006), pp. 1018-1027.

33. Frederick J. Zimmerman, Gwen M. Glew, Dimitri A. Christakis, Wayne Katon, "Early Cognitive Stimulation, Emotional Support, and Television Watching," *Archives of Pediatric and Adolescent Medicine,* vol.159, (2005), pp. 384-388.

34. James E. Burroughs, L. J. Shrum and Aric Rindfleisch, "Does television viewing promote materialism?" In: S.M. Broniarczyk and K. Nakamoto, Eds., *Advances in Consumer Research,* (Valdosta, CA: Association for Consumer Research, 2002), p. 1.

35. No author, "Cosmetic Plastic Surgery Research," (New York: *American Society for Aesthetic Plastic Surgery,* 2011), p. 1.

36. No author, "Beauty, Cosmetic and Fragrance Stores in the US: Market Research Report," *IBISWorld,* (March 2012), p. 1.

37. Rebekah Richards, "Gender Construction in a Ford F-150 Commercial," *Anthropology,* (November 7, 2009), p. 1.

38. Michael S. Kimmel, "Masculinity as Homophobia," In: Paul Murphy et. al., *Feminism and Masculinities,* (New York: Oxford University Press, 2004), pp. 182-190.

39. Lynn T. Lovdal, "Sex Role Messages in Television Commercials: an Update," *Sex Roles,* vol. 21, (November 1989), pp. 715-724.

40. Andrew Wernick, *Promotional Culture: Advertising, Ideology and Symbolic Expression,* (London: Sage, 1992).

41. Brian Baker, *Masculinity in Fiction and Film: Representing Men in Popular Genres,* (London: Continuum, 2006), p. 68.

42. U.S. Bureau of Labor Statistics, *Occupational Outlook Handbook,* (Washington, DC, 2012), p.1.

43.Lisa de Moraes, "Super Bowl XLVI: Biggest TV audience ever," *The Washington Post,* (February 6, 2012).

44.Doug Lesmerises, "In Sports, NFL Means Business," *The Wilmington News Journal,* (January 27, 2003), p. S1.

45. No author, "What is the NABC and What Does it Do?" *NABC,* (Kansas City, MO: no date), p. 1.

46. Mark Bradley, "Out of Control: Behavior of Football Fans Takes and Ugly Turn," *Atlanta Journal Constitution,* (November 6, 2002), p. 11.

47. Jamie Kritzer, "Are you Ready for some Football?" *Greenboro News and Record,* (September 12, 2002), City Life, p. 2.

48. Stanley J. Baran, *Introduction to Mass Communication,* (New York: McGraw-Hill, 2001), p. 280.

49. Gerald Mast, *A Short History of the Movies,* (Chicago: The University of Chicago Press, 1981), p. 260.

50. David Hawkridge and John Robinson, *Organizing Educational Broadcasting,* (Paris, UNESCO, 1982), p. 25.

51. Paul C. Beisenherz, "A Comparison of the Quality and Sequence of Television and Classroom Science," *Journal of Research in Science Teaching,* vol.10, (1973), pp. 355-363.

52. George N. Gordon, *Classroom Television,* New York: Hastings House, (1971).

53. B. J. Barth et. al. "Television Used in the Teaching of Science," *School Science and Mathematics,* vol. 58, (1958), pp. 202-204.

54. Edward Carlson and R. Wald, "Einstein's Universe," *Aviation and Space,* vol. 6, (1979), pp. 29-30.

55. U.S. Bureau of Labor Statistics, (Washington DC: *Occupational Employment and Wages,* 2012), p.27.

Chapter Eight

The Motion Picture

I.

No one person can be credited with inventing the motion picture. Instead we can record the names of some of the contributors to this revolutionary invention. Included is Thomas A. Edison, who projected a brief motion picture in New York in April of 1896. Edison followed the German inventors Max and Emil Skladanowsky, who demonstrated a "bioscope" in Berlin in November of 1895. In December of that same year, the Lumiere brothers produced a successful movie performance in Paris, France, even as Birt Acres, an Englishman, filmed the Oxford-Cambridge boat race, also in 1895. All of these "movies" consisted of a succession of photographs which were projected rapidly so as to create the illusion of movement on the screen.[1]

From these primitive beginnings came the silent films, the "talkies," colored movies, and television, all of which revolutionized American life. That is to say that motion pictures shaped the manner in which generations of Americans have understood their identity and even shaped the meaning of their lives. In fact, the movies have determined popular understanding of daily events because they have controlled the minds of millions during the twentieth and early twenty-first centuries. The motion picture has created the mass, which is one form of collective behavior. We define the mass as those anonymous people all affected by the same stimulus without knowing one another. By creating the repetition of the same stimulus over and over again in all parts of the country and even the world, movies have created reality for many millions who became addicted to this entertainment. Millions of moviegoers and television watchers have formed opinions of events based solely on the screen without really knowing what the events they watch in their living room are about. This ignorance is largely due to the failure of the vast majority of Americans to understand how movie images are constructed in the first place.[2]

The movies have political implications. Over the century that movies have grown from minor amateurish picture shows, movies have influenced the political landscape. In 2012, the motion picture industry is viewed as "liberal" on the grounds that the movies propagate in favor of labor unions, the Democratic Party, the "rights" of minorities, as well as the denunciation of Jews and the "rights" of women.

In this connection it is noteworthy that "Women Make Movies" was developed so as to deal with the near absence of women from the movie industry. "Women Make Movies" is a non-profit organization founded in 1972 in New York City. Its purpose is to allow women to learn the profession of movie director so as to overcome the almost total dominance of men in that field. Despite that effort, women still are only ten percent of American filmmakers. Men direct nine out of ten films, and on screen men outnumber women 77% to 17 %. Women also comprise only 17% of executive producers, producers, writers, cinematographers and editors. More than 20 percent of films released in the United States each year employ no women. This disparity has not improved in the years since 1990. In fact, in some of these categories, women are even less often employed than heretofore. Only 11% of film writers are women and 83% of the top 250 films had no women writers. Women are only 1% of cinematographers.

During the "silent" era of movie production, a good number of films were produced which dealt with immigration, the plight of sweatshop workers, labor unions, and a host of social problems. Yet, during the Great Depression, movies promoted the views of employers relative to the ongoing conflict between organized labor and capital.[3]

The messages that movies portrayed were and are so powerful because almost all Americans are consumers of these pictures. This has been the case for nearly a century. By 1920, one half of all Americans attended movie productions and by 1930 this had risen to one hundred percent of the population.[4]

Movies also influence the view Americans have of gender issues. Reviews of movies concerning the roles of women and men indicate that women have traditionally been portrayed as needing protection by well-intentioned men rather than adults capable of solving their problems themselves. Foreign, non-Anglo women are portrayed in movies as being sexually loose or lascivious.[5]

In recent years movies have portrayed labor unions and collective action by workers in
a negative light. This despite such movies as Coal Miner's Daughter, Saturday Night Fever, andRocky, all of which deal with individual lives but not the working class as a whole. In fact, such movies as Hoffa and Blue Collar seek to link unions to organized crime, a reputation well deserved.

II.

Movies are a business seeking to make a profit by attracting a sufficient number of customers. It is also a source of considerable employment involving a number of disparate occupations. According to the Bureau of Labor Statistics, 362,000 people were employed in the movie industry in 2008. Forty-four percent of these employees were working in management, business, and financial occupations; top executives constituted 12.3 percent, and advertising, marketing, public relations, and sales accounted for 3.2 percent of those working in the movie industry. Only eleven percent of those working in the movie industry were actors in 2008, while service occupations and professionals such as writers, cameramen, various technicians, and artists made up the other thirty percent of employees.

Although there are some "superstars" among actors who earn millions, the average income of actors is low. The Bureau of Labor Statistics shows that actors earn only $28.72 an hour or less than $60,000 a year if fully employed. However, acting jobs are intermittent, so that most actors must earn extra income from waiting on tables or driving cabs, etc. Producers and directors can earn more than $200,000 a year while camera operators earn a median salary of $41,000. In sum, it is evident that with the exception of the highest paid executives and actors, the motion picture industry is no road to riches.[6]

Income for movie actors and writers is also dependent on age and sex. This has been called "double jeopardy" because it affects women more than men. Not only are women not as often employed in the movie industry as is true of men, women are also refused movie roles as soon as they show the consequences of middle age. This is not true of men. In fact, middle aged men are given far more consideration for employment than not so young women. Furthermore, the Screen Actors Guild found that men earn more than women as movie actors. This in turn has consequences for American society as a whole as movie actors become role models for American society. Therefore, rejection of women by reason of age contributes to ageism in American society generally.[7]

In the acting profession there is no continuous employment. Acting is sporadic in that considerable intervals can occur as actors are chosen for a role in a film or theater production. This is true of men and women but affects women more, because the number of female roles in movies is smaller than the number of male roles. In fact, there are some films which have no women included but only men.[8] In 1948, the Supreme Court ruled that major movie studios could not own theaters. Then, too, television became more and more popular. This led the studios to cancel their long term contracts with movie "stars" and others, so that employment became more and more uncertain. Contracts were thereafter let for single films, so that it became possible to only hire younger actresses for every project.The obsession with younger actors and actresses also entered the television industry, because producers believed that they could not get their money back and make a profit unless they provided the audiences with the youngest possible actors.[9]

According to the United States Census of 2010, 45.3% of Americans were over age forty in 201,0 yet women over age forty received only 24% of movie roles, so that the word "older" refers to women older than thirty and men older than forty. Best actor winners are also more likely to be men of any age, while women who have been given that distinction tend to be younger.

The movie industry produces billions of dollars every year. In 2011 ticket sales reached $1.26 billion, a sum exceeded in every year since 1997, when $1.42 billion in tickets were sold. Even more was sold in 2002, when ticket sales reached $1.58 billion. In 1998, the movie Titanic alone earned $745,795,916, which was more than any movie earned between 1998 and 2011, when Harry Potter sold $381,011,219 in tickets.

Among movie distributors, Warner Brothers grossed 15.26% of the market between 1995 and 2011, followed by Walt Disney Pictures with 14.26%. Adventure movies earned more than any other genre with over 20% of the income.[10]

Like any business, the movie industry seeks to attract as many customers as possible. It is therefore not surprising that movies reflect the most popular tastes and opinions of moviegoers over the years. In the early years of movie production, the "working class" determined the contents of the primitive films then produced. After the First World War, this changed, as greater numbers of middle class and upper class citizens visited movie theaters. These new movie theaters were located in the suburbs and other wealthier neighborhoods than heretofore. Many of these theaters were literally movie "palaces," showing films that depicted lavish lifestyles and inducing the viewers to enjoy fantasies about riches and elegant sex.[11]

The change from appealing to the "working class" to depicting wealth was also induced by the censorship system that applied to the American movie industry. These censors were "self regulators," in that they were employed by the movie industry to follow a code adopted to forestall censorship by politicians.

Yet, these censors were in a position to ruin any movie producer they disliked. Therefore, movie makers produced only what pleased the censors and avoided sex, crime, violence, and any movie "calculated to stir up" antagonism between unions and capitalists. The studios wanted to make money and were not interested in fighting the censors at great expense while risking their investments.

For these reasons, changes were made in such feature films as "How Green Was My Valley" to remove scenes showing labor fighting mine owners. Likewise, scenes of a sit-down-strike were removed from An American Romance.[12]

The avoidance of working class experiences was transferred from the movie industry to television. In that medium too, the poor and certainly union members are portrayed as drunks, as people who don't work, as clumsy, uneducated, and incapable of any kid of leadership. Therefore, movies have a great deal of influence on popular opinions and therefore on politics, since large numbers of Americans gain their views of the world from the movies and nothing else. This then translates into votes and other expressions of popular culture.[13]

Although films are not to be confused with reality, they nevertheless mirror reality or "cause an illusion of reality." Evidently, if spectators could not discern the difference between reality and its representation in a film, they could not very well remain in their seats as horrible violence is depicted or "buffaloes stampede toward us." Nevertheless, movies are the dominant art form of the 20th and 21st centuries because movies grip us with more intensity than any book and because movies are so much easier to access. In order to read a book, one has to learn how to read. Movies require no such preparation. In order to understand a book, one has to know the vocabulary used by the author. This too is not needed when watching a movie. Furthermore, books require that the reader use his own imagination to "see" the scenes, the people and the actions in a book in his "mind's eye." This is also not required when watching a movie, as films provide all of these prerequisites, ipso facto.[14]

Movies are easily understood by a mass audience because pictures are easily understood and require no special learning, as is true of written works. Therefore, people who do not know English, are not scientifically sophisticated, and do not own any device common in industrial societies can nevertheless "enjoy" or at least appreciate a movie.It is for this reason that foreigners who may know only a few phrases of English import American movies. Movies are also easily accessible because the viewers are acquainted with the objects and symbols exhibited. What is true of foreigners, that is, non-English speaking audiences, is also true of children, who can understand a movie far faster than they can read a book.

Movies also are more absorbing than books. They compel attention and therefore eliminate outside distractions while in progress. That is certainly not true of books. Clever filming techniques such as moving the camera close to an object or further away or framing an object also promotes audience attention to the object at hand. As a consequence, people of all backgrounds can go to the movies and "see" the same thing. Both sexes, all races, all occupations, and viewers of every age can look at the same movies, albeit they give each movie an interpretation based on the viewer's own experiences. This phenomenon is called "selective perception," and occurs daily among almost all people, as we "see" what we want to see and never notice what doesn't fit.

III.

The influence of the movie industry on American culture is so strong because movies concerning almost every aspect of American life can reach the entire spectrum of people interested in a large variety of experiences. At least fourteen kinds of movie "genres" can be identified. These include military or war movies, movies dealing with sex, either as romantic fantasies or as "X" rated caricatures, comedies, crime stories, documentaries, dramas with historical overtones, so-called family movies, foreign films involving sub-titles, horror shows with utterly unrealistic scenes, science fiction movies, sports stories with particular emphasis on football, suspense stories such as produced by the late Alfred Hitchcock, "action" movies,and teen movies.

Perhaps the most successful war movie ever is Schindler's List, directed by Steven Spielberg in 1993. This movie earned seven "Oscars" as best movie of the year. The movie is based on the real life story of the Austrian businessman Oskar Schindler, who sought to make a profit from using Jewish slave labor in his factory during the Second World War. Instead of becoming an exploiter of Jewish suffering at the hands of the Nazi regime, Schindler saved the lives of 1,100 Jews from the gas chambers, only to die in poverty after the war.

A totally different war movie, equally successful, is the 1998 film Saving Private Ryan. Directed by Steven Spielberg, it was awarded five "Oscars." The movie deals with the invasion of Normandy by American forces on June 6, 1944, and moves on to follow the exploits of seven soldiers seeking to find a paratrooper who is the last surviving brother of four servicemen. The opening scenes, lasting 27 minutes, have been judged so realistic that men who were real veterans of D-day often left the theater because they could not tolerate reliving those gruesome experiences. The movie grossed $481.8 million worldwide.

Movies called "action" films generally include a great deal of violence and are therefore popular. The Dark Knight is an example of an action movie. Released in 2008, it grossed more than $533,000,000 in North America and over $1 billion worldwide. It received two Academy Awards. This movie is based on a comic book character and is pure fantasy. It appealed to a large audience because it contains a great deal of fighting and bloodshed. It has no other significance.

Likewise, The Lord of the Rings, directed by Peter Jackson, is a movie relying entirely on an imagined world of violent characters. The "plot" is nonsense but involves sufficient violence to produce over a billion dollars in the United States and $2.9 billion worldwide.This movie is one of three labeled a "trilogy," and appeals to anyone seeking to escape the real world for two hours.

Crime movies are immensely popular in the United States, although these movies have little to do with the reality of crime. In addition, the television audience is constantly entertained by the exhibition of criminal trials and scenes from real prisons. Fascination with deviant behavior is related to the need to segregate "the bad" from oneself so as to enjoy the superiority of the "in group" over the "out group." It is popular to pretend that the world is divided into the "criminals" and the "law abiding," although such a dichotomy is pure fantasy. In fact, it can easily be demonstrated that so called "authorities" include all kinds of criminals, while many of the so called "criminals" are among the 600 thousand convicted and imprisoned innocent Americans.

Two examples of "crime" movies which have been most popular and earned a great deal of money are "The Godfather" and "The Shawshank Redemption." "The Godfather" was a book written by Mario Puzo in 1969. The movie was directed by Francis Ford Coppola and starred Marlon Brando. The movie depicts ten years in the activities of an Italian-American crime family. Two sequences followed. This movie includes a number of disgusting murders and an immense amount of violence. It is nevertheless regarded as one of the greatest films ever produced. This violence is so popular that in 2006 a video game called The Godfather was produced. The first of the Godfather films grossed $246 million worldwide. In view of inflation, this cannot be compared to later movie successes. The movie won three Academy Awards and five Golden Globe awards.

The Shawshank Redemption did not become a huge success when first exhibited in 1994. However, it was subsequently revived on cable TV, and on tapes such as VHS, DVD, and Blu-rays. The movie was directed by Frank Darabont and deals with prison life, a corrupt warden, and resourceful prisoners who finally escape and start a new life. Nominated for several awards, the film received none. Yet, it continues to be popular. It was voted the favorite film by BBC viewers and listed as the favorite film of all time by BBC radio listeners.

Other winners in the competition for the dollar of the movie fans have been "Super 8," which deals with such issues as "aliens" from outer space and "zombies" as well as "spaceships." Released in 2011, this nonsense earned $127 million in the United States alone, and $260 million worldwide.

In 2009, the movie Avatar earned $2,782,275,172 worldwide. This movie deals with science fiction and depicts "events" in the 22nd century and later. It has no foundation in reality. No other movie has so far surpassed the $2 billion mark, although nine movies have earned more than one billion dollars. Except for Titanic, all of these "successes" are not based on any reality, although even Titanic is only vaguely related to any real events.

Despite the large income derived from a few movies, there has been a decided decline in the box office statistics in 2011. In fact, ticket sales in 2011 were about $500 millionless than in 2010. This is not a catastrophic sum compared to the total income from movie ticket sales in 2011, which has reached $10.1 billion. The fall off from 2010 is therefore less than 5%. Nevertheless, the income decline must be taken seriously by the movie industry, because that income includes revenues from more expensive 3D films. Furthermore, tickets now cost an average of $7.89. This has led to a decline in attendance by 5.3% since 2010, when attendance declined 6%.

The movies presented to the public during 2011 included "Mission Impossible," "Sherlock Holmes," "Alvin and the Chipmunks," and "The Girl with the Dragon Tattoo." These productions are so dull and uninteresting that fewer and fewer Americans want to subject themselves to the boredom. In part, the poor economy in 2011 contributed to this decline. In addition, a number of movies "bombed" because their content is so dull and unrealistic that even teenagers attend in lesser numbers than ever before. All kinds of other leisure activities compete with movies, and many people choose these leisure activities before they are willing to watch a movie.[15]

Movies have been described as reflecting conditions in the society. Yet, movies are also influential in their own right in determining opinions concerning manners and morals and beliefs common in American society. Because the first movies were shown in the immigrant slums of growing American cities, middle class "puritans" viewed such entertainment with anxiety and suspicion. Middle and upper class people were visiting "legitimate" theaters, and attending operas and ballets, and viewed movies as a dangerous preoccupation of the "lower" classes. In fact, the dispute over the movies was a reflection of the class warfare between labor and capitalists in the late nineteenth and early twentieth centuries. That dispute gradually diminished as the movies conquered the middle class in the United States, and theaters became commonplace in so-called "better" neighborhoods. As the nickel movie declined and long dramatic stories were shown in theaters, the cost of tickets rose to $2, a sum comparable to $22 in 2012 money.[16]

IV.

One outcome of the invention of the movie industry was that for the first time in American history, a group of "outsiders" gained influence on American culture. These outsiders were largely European Jewish immigrants who had arrived in this country between 1881 and 1923. They came from the Russian Empire because Czar Alexander III, the son of the assassinated Alexander II, promoted a violent persecution of the Jews in his empire, which included Poland and other eastern European countries. These immigrants began their American lives working in the garment industry sweatshops in New York and other large cities. For example, Adolf Zukor, later the head of Paramount, started as a sweeper in a New York fur factory. Likewise, Harry Cohn, later the head of Columbia, had been a fur salesman, and Samuel Goldfish, known as Goldwyn of Metro Goldwyn Mayer, became a glove cutter on arrival in the United States. Carl Laemle, head of Universal, worked as a clothing salesman, while William Fox of Fox Films worked twelve hours a day in a sweatshop beginning at age twelve. Louis B. Mayer sold junk.These men and others became Hollywood's "founding fathers" because the risky movie business could not attract established American born businessmen. Immigrants who had nothing to lose entered these enterprises while the native born and particularly the wealthy were not willing to be associated with the entertainment industry viewed as "lowbrow" or worse.[17]

These Jewish immigrants began their movie careers by owning a few nickelodeons, so named because the admission was five cents. From there they advanced to control complete monopolies, including production, distribution, and exhibition facilities. This success was made possible by a great deal of personal "drive," by willingness to work extremely hard, and by an understanding what the public wanted to see.[18]

V.

There was a time when movies were censored. The prime purpose of that self-imposed censorship was to keep any hint of sexuality from the viewers. The effort to censor movies consists at present of labeling movies R for those restricted to adults on the grounds they include sexuality, labeling G those films suitable for any audience, or PG those that require parental guidance.PG 13 are movies rated to be offensive to anyone under age 13 and NC 17 is a rating that prohibits children altogether.

Censorship of movies has been constant since movies were first produced. As early as 1909, New York City created the New York Board of Motion Picture Censorship. By 1915, this board had expanded nationwide and changed its name to "The National Board of Review." The censorship craze continued, so that by 1922 the Motion Picture Producers and Distributors of America, fearing the interference of Congress, hired Will H. Hays, former U.S. Postmaster General, as president. At a salary of $150,000 a year, Hays censored films for the next 23 years.. The principal purpose of this censorship was to eliminate any hint of sexuality from movies and to also prevent the exposition of movies which empathized with labor unions or dealt with the fate of the unemployed or the poor.[19]

The vehicle used to enforce movie censorship was the Production Code Administration, which "rigidly dictated the content of American films."Administered by Joseph Breen, this code banned words regarded as obscene and used commonly by such major authors as Ernest Hemingway in A Farewell to Arms.[20]

The censors, mainly recruited from Catholic self appointed moralists, claimed that "the picture people" had no decency. That was a code phrase for Jews, who were the target of the usual anti-Jewish canards taught in every Catholic church at the time. Catholics had formed the "Legion of Decency," which claimed to protect the public from the evils of sexy movies. Such movies as "Belle of the Nineties" were rejected by the New York censors and others because it depicted the actress Mae West in a tight fitting gown. Likewise, "The Awful Truth" was censored because the dialogue between Cary Grant and Irene Dunn could be interpreted as hinting at sexuality.[21]

Social problems were also censored in the early days of movie production. The censorship particularly targeted any movie depicting union activities or displaying the miserable life of mine workers or sweatshops. The effort on the part of the censors was to avoid angering big business owners. An excellent example of this kind of censorship refers to Black Fury, a movie produced in 1935 and dealing with the horrors of coal mining before any safety measures had been introduced into the mining occupations. All this censoring activity was resented by movie actors, producers, executives, and newspaper movie reviewers one of whom labeled the censor "the Hitler of Hollywood."[22]

Since the relaxation of these censorial practices, movie producers have made a large number of R rated movies which depict sexuality in a most graphic manner. No doubt this tactic brought in large profits when first used. Yet, at the beginning of the 21st century, R rated movies seem to bore the public, as ticket sales for these movies have slowed considerably since 2000.[23]

It is therefore peculiar that the movie industry, which exists to make money, would produce films that cost them money. Michael Medved argues that the actors, movie producers, and executives are motivated by the need to gain "insider" praise by making movies that do not "pander" to the public.[24]

Another reason for making numerous "R" rated movies is the need to sell American movies to a global audience. Sex sells and is appealing everywhere. Furthermore, American films, wich at one time reflected purely American culture, now appeal to worldwide cultures because the electronic age has made ethnic isolation obsolete. Hollywood in the 21st century has attained a global appeal. Likewise, movie producers in other countries, i.e. Brazil, Hong Kong, and India, have also "gone Hollywood" in that they seek to appeal to audiences worldwide using the same techniques once attributed to California based studios only. Movie producers therefore aim at the "average" movie fan worldwide. Movies, to be profitable, need to allow the audience to see part of themselves in them because the movie depicts scenes and events familiar to audiences everywhere.[25]

There are movies like Titanic which have had great success by attracting huge audiences around the world. Such movies have international audiences because they depict the human condition, which exists in all cultures and in all countries. Such movies are made mostly in the U.S.A. because the United States has the resources and the human skills needed to produce complex films. In addition, the United States is a relatively "free" country and is far less regulated by government. This allows movie makers more freedom to produce all movies generally censored in most countries in this world.[26]

Furthermore, the United States has a workforce educated in the production of films. There are universities in the United States which offer degrees in various aspects of moviemaking. These include New York University, University of Southern California at Los Angeles, American University, The American Film Institute, California Institute of the Arts, Columbia University School of the Arts, The North Carolina School of the Arts, San Francisco State University, University of Texas at Austin, University of California, and University of California at Los Angeles.

VI.

In the early years of movie production, European films not only comprised the majority of films shown in Europe, but before the First World War were also the majority of films shown in the United States. In 1900 the European film industry was dominant. At that time, the European share of the movie market reached sixty percent, mostly held by French companies but also by some German, Italian, and Danish producers.[27]

Twenty years later this had changed. Now European film producers sold their foreign subsidiaries, notably the Germans sold their American subsidiaries, to local film producers, and even sold some of their home shares to Hollywood. The First World War disrupted European film production even as the American movie industry expanded geometrically. After the war, actors were far more willing to come to Hollywood than to stay in Europe. Alone from the

Scandinavian countries, more than sixty actors and actresses migrated to the U.S. between 1914 and 1920 to gain fame and recognition far above what Europe had to offer. Only thirteen years later, the Nazi regime gained dominance in Germany. This led 470 of the most skilled, creative, and technically competent film producers to flee to Hollywood. Subsequently, the European film industry was used as a propaganda tool by the Nazi hierarchy and suffered a total collapse in the many countries occupied by Germany for the six years from 1939 to 1945.[28]

Since then, American movie production has been altered radically, as the great studios disappeared, television became dominant, and foreign film producers outside the western industrial countries competed successfully in the movie industry.

Although Hollywood, California, is still regarded as the headquarters of movie production in the world, there are a number of other countries which today (2012) outpace the United States in this regard. India is today the largest producer of films in the world. These films are made in the Hindi language and are most popular globally, even among people who do not understand the language. Because these films rival American productions, they are popularly called "Bollywood," a reference to Bombay, where most Indian films are produced.[29]

Few Americans have ever seen a "Bollywoood" film or heard of India's most famous movie star Shah Rukh Khan. As the most popular actor in Indian movies, he is regarded as a god or superman wherever in India he appears. The film industry where Khan appears is the most productive in the world. While the United States produces about six hundred films a year, Indian produces more than eleven hundred films a year. Indian films are popular in Asia, Africa, and South America. These films are not well received by American audiences because they seem juvenile and unsophisticated to Americans. In Indian films, which last as much as three hours, the stars sing and dance, change costumes several times, and participate in stories so utterly improbable that American audiences lose interest despite the indisputable fact that American films are seldom more than utter nonsense.[30]

VII.

Some nonsense movies are usually labeled "adult," obscene, or pornographic films, and are so large a segment of film production in the United States that they cannot be overlooked concerning the influence of movies on American culture. While Hollywood produces about six hundred films a year, the adult industry produces nearly 13,000 films a year. The sex based industry generates an estimated $13 billion a year. This includes all kinds of products other than films, which reportedly earn only $980 million annually.[31]

Adult films may be divided into "reel films" and the "video" era, which began in the 1980's. These films became a good deal more explicit than was true of the earlier reel films and depict a variety of sexual situations. By the 21st century, the studios which make these films have adopted a number of techniques used in the normal feature films produced in Hollywood but also in a number of other locations around the world. Adult films have now adopted mainstream media techniques, as they are now producing digital films.[32]

Although the "video era" made adult films more accessible to large audiences than was ever the case in the days of the "reel" films, the year 1972 was a landmark in the production of these films because in that year "Deep Throat" became the first widely accepted legal narrative film in the adult genre.[33]

There were several reasons for the emergence of these "hard core" films in the 1970's and the 1980's. One of these was the development of 16mm film, which allowed pornographers to compete directly with other filmmakers. In addition, pornographic films had a good deal of influence on the entire film industry, in that the censorship of pornography was eased, so that producers of mainstream films began to bring sexually explicit scenes to the screens of their productions. The easing of the censorship codes came about because the studio system crumbled in the 1950's and innumerable private individuals used the 16mm camera to produce pornography, which was then shown in improvised theaters around the country. Finally, pornographers exhibited their wares in theaters and took advantage of a number of Supreme Court decisions, like the Roth Decision, which favored them and their interests.[34]

The "Sexual Revolution" of the 1960-1980 decades also contributed a great deal to the acceptance of pornography among moviegoers in the United States. This led to the introduction of women into the audiences of these pornographic films. Before the 1960's, sexual revolution "sexploitation" films were seen almost only by men. By 1972, couples were in frequent attendance at these showings, so that by 1973, theater owners estimated that 80% of those viewing pornographic films were couples. Moreover, theaters exhibiting pornography were earning in the vicinity of $40,000 weekly at that time.[35]

In the United States, pornography is considered a form of personal expression and is therefore protected by the First Amendment to the U.S. Constitution unless it is obscene. The First Amendment states in part: "...or abridging freedom of speech or of the press etc." Rulings concerning pornography and obscenity began in 1897 with Dunlap v. U.S. in which the court upheld the conviction of the owners of a newspaper for mailing "obscene, lewd, lascivious and indecent material" and have continued ever since, reaching over thirty decisions both in favor and opposed to censorship and leading to utter confusion as to what is or is not obscene. Among other decisions, the Roth decision of 1957 stands out because it has been repeatedly cited as the landmark decision concerning obscenity.In that case the court ruled that obscenity is not constitutionally protected. The court ruled further that the standard for judging obscenity is "whether, the average person applying contemporary community standards (believes) it appeals to prurient interests." [36]

While the Roth decision and others have secured the right of an adult to view pornographic materials, the Supreme Court has upheld laws prohibiting child obscenity and pornography. In 1982 the court held in New York v. Ferber that child pornography is not protected by the constitution.

VIII.

Although the most famous actors earn millions, the average actor who appears in two films a year earns only about $24,000 a year. That low income is principally the result of the intermittent work available to actors. For example, a little known actor, Dylan Baker, made five guest appearances and worked in two films in 2010. He earned $8,000 a week and worked three weeks. By appearing in some TV shows he made another $12,000 for a total annual income of $36,000. Even those actors who earn more must still pay their agent ten percent, their lawyer five percent, about 12 percent to a manager, and $4,000 to a publicist. The Screen Actors Guild, the union of screen actors, requires that members earn at least $2,352 for a five day week. The Screen Actors Guild has about 100,000 members, of whom only about fifty may be considered "stars." Indeed, these "stars" earn millions each year. Yet, the average income of the vast majority of screen actors is less than $5,000 a year since employment is erratic.[37]

Despite the poor prospects of gaining employment, there are always those willing to train for the approximately 155.000 jobs available to actors, producers, and directors in the motion picture, video, and broadcast industries. The most common way of entering into the profession is to attain a college degree in the theater arts and sciences. The National Association of Schools of Theater accredits over 150 programs in theater arts. Graduates of such programs have to compete with the children and other relatives of established members of this occupation, who are preferred and generally succeed in gaining the most lucrative positions in acting and other aspects of the industry.[38]

Examples are Angelina Jolie Voight, daughter of actor Jon Voight, Peter Doocy, son of anchor reporter Steve Doocy on Fox News, Miley Cyrus, daughter of singer Billy Ray Cyrus, Liza Minnelli, daughter of actress and singer Judy Garland, Anne Lockhart, daughter of actress June Lockhart and granddaughter of actors Gene and Kathleen Lockhart, Isabella Rossellini, the daughter of actress Ingrid Bergman, Ben Stiller, the son of actors Jerry Stiller and Anne Meara, Barry van Dyke, the son of actor Dick van Dyke, and Stephanie Zimbalist and Efrem Zimbalist, Jr.,the children of violinist and composer Efrem Zimbalist, Sr,

A much longer list of such relations can be produced, showing that the obstacles to gaining access to the acting profession are far greater than the need for talent and education.

Many actors leave the profession after a short time because the competition for work is so great, the pay so low, and the work exceedingly hard. These facts conflict with the myths concerning the acting profession, which view acting as glamorous and lucrative and invariably the road to fame. The truth is evidently quire otherwise.

Therefore, there is a considerable gap between the reality of working in the movie industry and the beliefs concerning such work. These beliefs have entered American culture not only because the movies visible to the public are far more exciting than the reality, but also because all opinions concerning occupations differ from their actual practice.

IX.

The influence of the movies on American culture are profound. This is true for the individual movie customer as well as American society as whole. For one, movies, whether seen in a theater or on the television screen, stimulate impulses which are ordinarily restrained. This is particularly true of aggression, normally repressed.

It would be a gross exaggeration to claim that all violent movies or video games must lead to aggressive conduct by the viewers. It is, however, visible that some people, particularly less experienced young folks, can be led to violent behavior in that their inhibitions may decline on viewing violence repeatedly. This is in part an explanation of the 1999 assault on students and teachers at Columbine High School in Littleton, Colorado, by Eric Harris and Dylan Klebold, two students at that school. The videos watched by Harris and Klebold involved bloody shooting scenes which both boys reenacted by making videotapes involving guns and the killing of school athletes. There are of course numerous video games, movies, television shows, and comic books which deal with extreme violence and which are accessible to anyone without respect to age.

Anderson and Dill published an in depth study of the consequences of screen violence on men and women. This study shows that "exposure to violent video games will increase aggressive behavior in both the short term and long term." Anderson and Dill show that some people are aroused by seeing violence on the screen and consequently attribute the anger provoked by such exhibitions to people in their vicinity, leading to violence against other persons.[39]

There are evidently people who have an aggressive personality and are therefore easily provoked to violence by viewing violence. Furthermore, Anderson and Dill found that some viewers of violence conclude that violent behavior is an acceptable means of solving problems. Craig Anderson also studied the effects of violent movies on aggressive thoughts, and reached similar conclusions as were found in connection with video games. Furthermore, there is a considerable literature confirming these findings over several decades.[40]

Movies have an immediate impact on viewers. This is particularly true of children, who do not easily distinguish between reality and "make believe." Prior to several decisions by the Supreme Court and the Court of Appeals, censors in numerous American jurisdictions prohibited children from entering movie theaters which exhibited violence or other activities protected by the first amendment to the constitution. Then, in 1965, the Interstate Circuit Court in Texas ruled, and the Court of Appeals affirmed, that even the community's interest in protecting children did not justify excluding children from contact with constitutionally protected expression.[41]

The movies are very much responsible for the view which American women and men have of themselves. Movies create "ideal types" which most Americans seek to imitate, although the actors who represent these "ideal types" are seldom what they seem. Max Weber, the prolific German sociologist of the early twentieth century, discussed the concept of "ideal type" at length without reaching any comprehensible conclusion. In American society generally, "ideal types" are those who become our role models, usually portrayed by movie actors. Accordingly, women have physical attributes represented by movie actresses, whose beautiful faces and glamorous bodies others seek to imitate. Many of these actors depend for their looks entirely on cosmetics and makeup artists because without such intervention most actors look as common as everyone else. Nevertheless, actors have so much influence on Americans that numerous magazines are sold at checkout counters in food stores and elsewhere which are devoted to the behavior and appearance of movie actors. Such magazines depict the antics of actors, and often use only the first names of the actors discussed because readers of these publications recognize the actors at once. This demonstrates that there are a good number of Americans who spend much time reading about and viewing movie actors in an effort to imitate them.

What is true of women is also true of men. Male actors generally portray physical strength, aggression, and toughness. These traits are then imitated by men, particularly young men, who seek to make an impression on women by exhibiting conduct approved by movies and their customers. Such actors as Sylvester Stallone and Clint Eastwood are favorite male role models, although hardly any American man can conduct his life according to the roles these actors portray and which are not representative of the actors' lives. A good example of the discrepancy between the act and the real behavior of an actor were the many portrayals of fighters and war heroes as depicted by John Wayne, who made thirteen war related movies in which he played military heroes. Yet, Wayne dodged the draft then in effect and never served in the armed forces.[42]

The movies portray a number of American values and attitudes which are reinforced dramatically by this medium. The "rugged individual" is a favorite movie theme, in which a strong willed individual overcomes all odds and succeeds. That success may be financial or military or occur in the sports arena. The heroes of such movies are democratic and demanding freedom of action for themselves and others. These too are American values. Individualism also involves the ambition to gain success at any cost and alone. The lone Western hero is a favorite American movie character who probably never existed but, portrayed by Clint Eastwood, reflects deeply held beliefs about self reliance and incentive in the context of the American experience. All this adds up to an achieving society, which is the core of American culture.

Every literate society includes three kinds of customs. Folkways, which carry no sanctions, laws, which are written statutes, and mores, which are expressions of public opinion and have considerable influence on human conduct, although mores are not enforced by law.

The movies and the actions of movie actors who are culture heroes and therefore role models influence American mores considerably. In fact, movies and the actors who appear in them have been instrumental in changing American behavior, although such influences operate over a long period of time, as mores change slowly.

An example of the influence of movies and actors on American mores was the so called "scandal" associated with the sexual union of actress Ingrid Bergman with Italian film producer Roberto Rossellini. Bergman, a native of Sweden, was married to a dentist when, in 1950, she traveled to Italy to make a film with director Roberto Rossellini. While there, and still married, she entered into an "affair" with Rossellini, who was also married at that time. When she became pregnant with their son, Renato, she unleashed a huge scandal in the United States. She was barred from appearing on the "Ed Sullivan Show," a popular variety television production. She was subsequently denounced on the floor of the U.S. Senate and therefore moved to Italy, leaving her husband and daughter Pia Lindstrom behind in Rochester, New York. Their divorce was widely publicized in the "Victorian" 1950's, and Bergman became a "persona non grata" on having twins with Rossellini in 1952. Viewed from the vantage of 2012, the scandal surrounding Bergman seems ridiculous, as American mores have changed radically since then. Millions of Americans are now divorced. Fifty one percent of American children were born to single mothers in 2011, and extramarital "affairs" are so common in 2012 that they are no longer subject to discussion. No doubt, influences other than the movies were also partially responsible for these changes in American mores. However, the influence of the movies and their actors on millions of moviegoers and television watchers cannot be denied. It is indeed profound.[43]

The movies also impact American education. There is hardly a school, from kindergarten to university, which does not have an 'audio-visual' library including both educational films and entertainment. The use of movies in schools began in 1932, when the University of Chicago released two ten minute educational films called "The Molecular Theory of Matter" and "Oxidation and Reduction." The idea was to reduce the time needed to discuss these topics from five to six hours in the classroom to twenty minutes. Since then, thousands of educational movies have been made and used in just about every area of education on every subject in every school.

X.

The foregoing has demonstrated that movies are indeed a powerful medium, transcending class, ethnicity, gender, religion, and age. The reasons for this power include the ability of movies to appeal to people who have no training or education in the movie business and who need not have been educated to know anything in particular to understand the message of any movie. Even small children can react to movies and react to views of objects and people with whom the child has become familiar. Unlike the need to be literate in order to read a book or to have some education in British literature to understand Shakespeare, no preparation of any kind is needed to understand a movie. Therefore, masses of people of all backgrounds can enjoy and react to the same movie. This is in part due to the ability of movie producers to exclude stimuli other than the present scene from the attention of the audience. That is not possible for the theater director.

Movies are not reality. Evidently, movies are not reality, since few if any could sit by quietly in a movie theater and watch bloodshed and mayhem or, for that matter, sexually arousing scenes. However, movies imitate reality sufficiently to allow almost anyone to become engrossed in the pictures portrayed and the story told, which usually deals with some aspect of human action.

Movies also teach various ideologies. For example, "Mr. Smith Goes to Washington," released in 1939, teaches the importance of democracy, even as in Europe, the Nazi government of Germany used propaganda movies to teach the viewers the cult of the "Führer," dealing with absolute obedience to the dictator.

Movies largely rely on posing some questions which are answered later in the show. For example: "Will the heroine fall in love with the scoundrel or the football hero?" or, "Will the detective, Columbo (Peter Falk) succeed in finding the killer?" or, "Will that soldier survive the horrendous attack by a heinous enemy?"The audience won't find out until the end of the movie, and is emotionally charged to remain in their seats and wait for the answers.

Of course, the influence of movies on Americans is also promoted by incessant advertising. Together with businesses of any kind, advertising is a constant feature of American life, so much so that much of it is "tuned out" by the mass audiences advertisements seek to influence. Nevertheless, the movies constantly advertise sex, action, excitement, new experiences, music, and beautiful pictures. In sum, it is certain that movies are one of America's most unique and important inventions, which revolutionized social life in this country and around the world. Yet, this was by no means the end of American ingenuity. Television and the internet created yet another new world, a world that never was but that has changed America profoundly.

Summary

Many inventors participated in creating the movies. Movies create mass stimuli, which may have political implications and are certain to impact the mores of Americans. Movies are primarily a business, which furnishes the owners with vast profits but allow actors a very small income, unless the actor is a superstar making millions.

There are many kinds of movies, including war movies, those labeled adventure movies as well as action films, crime movies, romances, biographies, educational movies, and many more. These movies were at one time censored by a board concerned with violence and sex. This led to the adoption of a rating scale by the movie industry, which also appointed a censor assigned the task of keeping children from seeing violence and adults form seeing sexual conduct.

India is the largest movie producer in the world, producing thousands of movies each year in the Hindi language.

In addition to so-called legitimate movies, there is also in the United States a pornographic movie industry, which has been ruled constitutionally protected according to the U.S. Supreme Court "Roth decision."

In sum, the influence of the movies on American culture is so great that the moral standards of some movie actors have become the standard for Americans generally, as such actors are the "role models" of Americans at the beginning of the 21st century.

Notes

1. C. W. Ceram, *The Archeology of the Cinema,* (London: Thomas and Hudson, 1965), pp. 157, 201.

2. Robert Sklar, "Oh Althusser! Histiography and the Rise of Cinema Studies," In: Robert Sklar and Charles Musser, Editors, *Resisting Images: Essays on Cinema and History,* (Philadelphia: Temple University Press, 1999), pp. 12-35.

3. Kay Sloan, *The Loud Silents: Origins of the Social Problem Film,* (Urbana, IL: The University of Illinois Press, 1988), pp. 75-76.

4. Steven J. Ross, *Working Class Hollywood: Silent Films and the Shaping of Class in America,* (Princeton, NJ: Princeton University Press, 1998), p. xiii.

5. Ben Singer, "Female Power in the Serial Queen Melodrama," *Camera Obscura* , vol. 22, (1990), pp. 91-129.

6. U. S. Department of Labor, Bureau of Labor Statistics, "Television, Video, Motion Pictures, etc.," *Occupational Outlook Handbook 2010-2011,* (Washington, DC: U.S. Government Printing Office), pp. 1-12.

7. Marilyn Zeitlin, "Too Old for Hollywood," *The Progressive,* vol. 50, no.1, (January 1992), pp. 33-34.

8. Susan Sontag, "The Double Standard of Aging," In: Juanita H. Williams, ed., *Psychology of Women,* (New York: Academic Press, 1970), pp. 462-478.

9. Doris G. Bazzini, "The Aging Woman in Popular Film," *Sex Roles,* vol. 36, pp. 531-543.

10. Nash Information Service, *The Numbers,* "U.S. Movie Market Summary 1995-2011," pp. 1-3.

11. David Nasaw, *Going Out: The Rise and Fall of Public Amusements,* (Cambridge: The Harvard University Press, 1993), pp. 220-240.

12. Steven J. Ross, *Working Class Hollywood,* (Princeton, NJ: Princeton University Press, 1999), pp.57, 116, 133, 198.

13. Miriam Hansen, *Babel and Babylon,* (Cambridge: The Harvard University Press, 1991), Chapter 3.

14. Noel Carroll, "The Power of Movies," *Daedalus,* v. 114, no.4, (Fall 1985), p. 80.

15. Brooks Barnes, "A Year of Disappointment at the Movie Box Office," *The New York Times,* (December 25, 2011), Business section, p. A:1.

16. Andrew Hacker, "The Hollywood Influence," *The New York Times,* (November 9, 1973), p. BR3.

17. Irving Howe, *World of Our Fathers,* (New York: Harcourt Brace Jovanovich, 1976), p. 164.

18. Neal Gabler, *An Empire of Their Own: How the Jews Invented Hollywood,* (New York: Crown Publishers, Inc., 1988), pp. 54-55.

19. Stephen Tropiano, *Obscene, Indecent, Immoral and Offensive: One Hundred Years of Censored, Banned and Controversial Films,* (New York: Limelight Editions, 2009), p. 51.

20. David Cook, *A History of Narrative Film,* (New York: Norton, 1981), pp.266-267.

21. Michael Foucault, *The History of Sexuality: An Introduction,* (New York: Vintage-Random, 1980), p. 24.

22. Glyn Roberts, "The Hitler of Hollywood," *Film Weekly,* (August 31, 1934), p. 11.

23. Michael Medved, *Hollywood vs. America: Popular Culture and the War on Traditional Values,* (New York: Harper Collins, 1992), p. 290.

24. Ibid., p. 306.

25. Bruce Bower, "Average Attraction: Psychologists Break Down the Essence of Physical Beauty," *Science News,* vol. 145, no. 12, p. 182.

26. Michael Porter, *The Competitive Advantage of Nations,* (New York: The Free Press, 1990), pp. 18-20, 64, 67-68, 114, 115, 146.

27. Ralph Mottram, "The Great Northern Film Company," *Film History,* vol. 2, no.1, (1988), pp. 77-89.

28. Harold J. Wollstein, *Strangers in Hollywood,* (London: The Scarecrow Press, 1994).

29. Nareen M. Kabir, *Bollywood: The Indian Cinema Story,* (London: Channel 4 Books, 2001), p.1.

30. Suketu Metha, "Welcome to Bollywood," *National Geographic,* vol. 207, no. 2, (February 2005), pp. 1-9.

31. Mark Kearns, "State of the Adult Film Industry," AVN Magazine, (March 2007).

32. Chauntelle Anne Tibbals, "From the Devil in Miss Jones to DMJ6-power, inequality, and Consistency in the Content of US Adult Films," *Sexualities,* vol. 2010 13:625.

33. Linda Williams, *Hard Core, Power, Pleasure and the 'Frenzy of the Visible',* (Berkeley: The University of California Press, 1999), p. 98.

34. Jim Holliday, "A History of Modern Pornographic Film and Video," In: James Elias ed., *(Porn 101: Eroticism, Pornography and the First Amendment,* (Amherst, NY: Prometheus, 1999), pp. 341-351.

35. Rex Williams, "The Porno Movie Scene," *Sensuous One,* (1973), p. 9.

36. Roth v. U.S., 354 U.S. 476 (1957)

37. Bureau of Labor Statistics, U.S. Department of Labor, "Earnings," *Occupational Outlook Handbook, 2010-2011 Edition,* (Washington DC: The United States Government Printing Office).

38. Bureau of Labor Statistics, U.S. Department of Labor, "Actors, Producers and Directors," *Occupational Outlook Handbook, 2010-2011 Edition,* (Washington DC: The United States Government Printing Office).

39. Craig A. Anderson and Karen E. Dill, "Video Games and Aggressive Thoughts, Feelings, and Behavior in the Laboratory and in Life," *Journal of Personality and Social Psychology,* vol. 78, no. 4, (2000), p. 772.

40. Craig Anderson, "Effect of Violent Movies and Trait Hostility on Hostile Feelings and Aggressive Thoughts," *Aggressive Behavior*, vol. 23, (1997), pp. 161-178.

41. *Interstate Circuit v. City of Dallas*, 249 F. Supp.19 (N.D. Tex. 1965).

42. Donald Shepherd and Robert Slatzer, *Duke: The Life and Times of John Wayne*, (New York: Kensington Publishers, 1985), pp. 50-51.

43. Charlotte Chandler, *Ingrid: Ingrid Bergman, A Personal Biography*, (New York: Simon and Schuster, 2007), pp. 149-209.

Chapter Nine

The Airplane

I.

Greek mythology records that a flying horse, Pegasus, and the messenger god Hermes flew through the air in ancient Greece. In addition, Daedalus made wings for himself and his son Icarus. These wings were made of feathers and wax, and let Icarus escape from captivity. However, Icarus flew too close to the sun, so that the wax melted and he was killed when the wings failed him. Daedalus also met with misfortune, as he flew too low and crashed into rocks, so that he fell and drowned. Likewise, a British king, Bladud, who ruled in the ninth century B.C.E., flew with wings he had made but was killed when he landed on top of the Temple of Apollo in the town of Trinovantum. Archytus of Tarantum also flew a wooden bird he had made about 400 B.C.E. Archytus was the founder of mathematical mechanics and was therefore equipped to design and build the first self propelled flying device. That device was propelled by a jet of what must have been steam. The device actually flew about 200 meters or 656 feet. The device was suspended on a wire or pivot. In fact, the Aristotelian Mechanical Problems was probably written by Archytus but misattributed.[1]

A number of ancient artists portrayed animals and humans with wings flying about the landscape. This was true among Romans and Greeks and others and reflects "wishful thinking," as none of the ancients knew aeronautics or the principles of attaining powered flight.

The Chinese were the first to fly kites sometime during the first millennium. These kites were used to carry bombs in war or were used in reconnaissance. The thirteenth century Italian explorer Marco Polo described how the Chinese had bound a man to a wooden kite, who was then carried up by the wind and flew briefly. Several Turks jumped from rooftops with wings attached to their bodies, but all died from these attempts at flying.[2]

The ancient Babylonians refer to flight in a set of laws called "Halkatha." In an epic called "Etana," the Babylonians described in cuneiform, dating back three thousand years, Etana's prehistoric flight. India also has a tradition of ancient flying machines. A sacred Hindu book called Samarangana Sutradhara contains over 200 references to flying. According to a recent book called Aeronautics: A Manuscript from the Pre-historic Past, the ancient Hindu book accurately deals with aerodynamics, machines, and equipment close to what is known today.[3]

In the 1480's, Leonardo da Vinci made a number of drawings of birds in flight as well as a possible flying machine for human flight. Nothing came of this, as no one made an effort to actually build a flying machine according to da Vinci's projections.[4]

In a number of European countries, various ideas favoring flight were proposed, but never succeeded until the Montgolfier brothers flew a balloon with a basket holding sheep and a duck in 1783. Later that same year, Étienne Montgolfier became the first human to lift off the earth and fly a short distance followed by Jean-François Pilâtre de Rozier, who flew at an altitude of about 80 ft. for a distance of about 3000 ft. These flights became a sensation in France and other countries, followed by numerous other launches. Most of the early balloons contained hydrogen. One of these hydrogen balloons was used to cross the English Channel in 1785.[5]

During the nineteenth century numerous balloon flights were recorded. Then, in 1891, Otto Lilienthal succeeded in building a hang glider that allowed him to sit in the glider and fly 24 meters or 79 feet. Lilienthal made a number of improvements to his hang gliders, but was killed in a glider crash in 1894.[6]

These experiments in flying heavier than air machines culminated in the successful sustained and controlled flight by Orville Wright on December 17, 1903 at Kitty Hawk, North Carolina.The Wright brothers, Wilbur and Orville, did not succeed suddenly by devising a plane that would fly, but achieved this gradually. Prior to 1903, the Wrights designed and built a wind tunnel and instruments which allowed them to predict the performance of their flying machine. These inventions made the first flight possible, even as today, in 2012, pressure testing machines and studies of wind velocity, atmospheric density, drag and variations in the angle of incline are still used to construct the modern airplane.[7]

According to engineering principles, there are five parts to an airplane. These are the wings, stabilizing and controlling surfaces, power, body, and landing gear. The Wright brothers recognized this because they had experimented for some time with glider planes but without power, as there was no engine which could be installed in their primitive invention. At first they sought an automobile engine which would fit their intentions, but found no useful engine. They therefore built their own four cylinder engine producing twelve horsepower. Without sufficient lubrication, this first airplane engine would not run more than a few minutes. In 1905, the Wright brothers offered their invention to the U.S. Army but were turned down. Even the patent office was skeptical, and spent three years, until 1906, to approve the "flying machine."

Then in 1909, Glenn H. Curtiss constructed an airplane with a fifty horsepower engine weighing 250 pounds. This achievement caused a number of European inventors to produce larger and more efficient engines and also made Curtiss the most successful airplane manufacturer in the world, headquartered in Buffalo, New York. [8]

II.

The First World War of 1914-1918 made the airplane a fighting instrument. The new aerial weapon was terribly primitive from the view of the 21st century. The planes were fragile and made of wood, wire, and fabric. This meant that the pilots were as much endangered by their own aircraft as that of their enemies. Furthermore, there were no known tactics or strategies for the use of airpower. Therefore, pilots and ground personnel had to devise such tactics and learn from experiments and often deadly mistakes. In order to reduce casualties, bombing raids were conducted at night so as to avoid antiaircraft fire. Because there was no radio communication, long range bombing raids against bridges, rail points, and ammunition dumps were difficult to coordinate, so that balloons were used for aerial reconnaissance.

In the course of the war, pilots on either side of the conflict constantly changed tactics to reflect changes made by the enemy. As the war continued, new equipment was developed by both sides. Aerial combat was so dangerous that the average aviator lived only three weeks.[9]

In addition to the land based air corps which fought in the European combat zone, the U.S. Navy brought about a new naval strategy based on air power. This all began in 1910, when civilian pilot Eugene Ely flew a Curtiss biplane from the deck of the cruiser U.S.S. Birmingham at Hampton Roads, Virginia.

Development of naval aviation was at first slowed by the firm belief of naval commands that sea power depends on battleships, as taught by Alfred Mahan, the most influential naval strategist of the nineteenth century whose book, The Influence of Sea Power Upon History 1660-1783 became "The Bible" of naval strategy since its first publication in 1890. Mahan showed in that book that those countries which had a navy had greater worldwide impact than those who did not have access to the sea.[10]

Naval aviation was not taken seriously by the brass running the navy until the possibility of a sea war with Japan became a reality in the 1920's. It was in that year that Congress funded the conversion of an existing naval ship into the first American aircraft carrier. The test for the use of aircraft carriers came during the Second World War, when at the Battle of the Coral Sea, American planes succeeded in halting the Japanese advance toward Australia. The second test of air power at sea was the Battle of Midway from June 4-7, 1942, in which American naval air power destroyed four Japanese aircraft carriers together with their most highly trained pilots.[11]

Nevertheless, it cannot be denied that the Japanese navy taught the U.S. navy a lesson concerning the use of naval aviation in war when they attacked

Pearl Harbor by air on December 7, 1941, and destroyed the entire Pacific fleet except for the aircraft carriers, which were not docked at Pearl Harbor on that fateful day, but were at sea.

As the Second World War progressed, air power became more and more important both to the Allies and to the German-Italian Axis. The rehearsal for the war which began with the German invasion of Poland on September 1, 1939, was the intervention of Germany in the Spanish Civil War of 1936. In 1937, when it appeared that General Francisco Franco had too few men and not enough war material to win the war against the Republican Spanish government, Hitler decided to help Franco. Therefore Germany sent the "Condor Legion" to Spain. The "Condor Legion" consisted of several squadrons of dive bombers and other planes, which not only succeeded in giving Franco victory but gave the German pilots experience in strafing and bombing targets. That experience was then activated in 1939, when dive bombers called "Stukas" or "Sturzkampfflugzeug" used a screeching siren attached to each plane to terrify the population targeted. This tactic was part of the German "Blitzkrieg," or lightning war, which led to the capitulation not only of Poland but nearly all countries invaded by the German armies in the first part of the Second World War. The lesson was learned by the Allies, England, the United States, and Canada and other participants in the effort to confront the Germans and the Japanese. The lesson was that air power would make a decided difference in the outcome of the war.[12]

For the United States, the need to increase the U.S. Air Force was evident. Consequently, the United States produced 325,000 planes, which was far more than Germany and Japan could produce together. These planes not only won the war in Europe by gaining air superiority over the German "Luftwaffe" (air weapon), but were also instrumental in delivering the atomic bomb on Hiroshima and Nagasaki, thereby forcing the surrender of Japan.[13]

III.

The use of the airplane by the military is of course not the only consequence of this revolutionary invention. One of the earliest uses of air transportation was the Air Mail service, which began in 1925 when the Kelly bill passed Congress, authorizing the private operation of air mail service under contract to the U.S. government. At the beginning, yearly air mileage was 599,244. This increased to 2,076,764 miles by the end of 1925, an increase of 275 percent. In terms of pounds of mail, the increase went from 526,578 pounds to 1,500,000 pounds, or 288 per cent.[14]

Despite these gains, few people had an interest in air transportation in 1925, so that the Springfield, Illinois postmaster, William Steiger, wrote that air travel was viewed with "public apathy." Evidently, the development of the airplane since its invention in 1903 was greatly enhanced by its use during the First World War. Thereafter, it was the U.S. government use of air mail which kept interest in air travel alive, particularly since government subsidies were needed to promote such aviation. A considerable effort was made to convince business firms to use air mail. This meant that an educational campaign was needed to introduce air mail to the public. An example of an early air mail service was the one inaugurated from Springfield, Illinois to Chicago and St. Louis on April 15, 1925 by Charles Lindbergh, later to become an American hero after he flew "The Spirit of St. Louis" from New York to Paris in 1927.[15]

By 1926 air mail extended from New York to Chicago and from there to Dallas and west to San Francisco, operated by a private firm called National Air Transport. This meant that one or more business days could be saved by using air mail, which cost somewhat more than "regular" mail and was not used very much in the 1920's and early 1930's. Furthermore, the planes used by the various private air mail services were not very safe. These planes did not have wheel brakes and flew only in daylight and were delayed by bad weather, particularly fog. Nevertheless, in February of 1934 a plane flew, for the first time, across the continent from Los Angeles, California, to Newark, New Jersey, and thereby reduced the total traveling time from coast to coast by six hours.[16]

The dangers associated with flying in the 1920's and 1930's were best illustrated when the U.S. Postmaster General rescinded all contracts for private airlines to fly the mail across the country. The Franklin Roosevelt administration claimed that a good deal of nepotism and particularly influence peddling corrupted the contract process.[17]

Whatever the merits of these accusations, in February of 1934 the task of flying the mail across the country was assigned to the Air Corps, the predecessor of the U.S. Air Force. That this was a fatal mistake became evident only after the air corps registered four times the fatal accident rate as commercial air lines had suffered. Between February 19, 1934 and May 8, 1934, there were sixty-six accidents, resulting in the deaths of 12 Army pilots. The reasons were not only that few air corps fliers had much experience in piloting the rather primitive planes of that day, but also that the air corps did not even have the safety devices then available. Few air corps planes had a radio. Ground control did not exist. There was no radar and navigational devices were hardly available. Furthermore, weather prediction was unreliable. This meant that commercial airline pilots, equally devoid of the devices now available, had fewer accidents than air corps pilots because they had more flying experience.[18]

On May 8, 1934, air mail was once more entrusted to commercial carriers who had contracts with the U.S. Post Office to deliver air mail. This continued until October 10, 1975, when all domestic first class mail was delivered by air wherever practical at the normal first class rate. On May 1, 1977, the Post Office Department became the U.S. Postal Service. In 1995, the USPS began sending all international First Class mails by air without additional charge, so that the phrase "air mail" was no longer used on any stamps. It is evident that the influence of the airplane on mail communications was the first and most effective manner in which the United States entered that age of aviation.[19]

In the 1940's, the image of the Air Corps changed from a failed and incompetent operation to a major contributor to the war effort and eventual victory over Germany and Japan. It was the U.S. Air Force, together with the British, who bombed German cities into ruins and who achieved final victory by delivering the atomic bomb on Nagasaki and Hiroshima in Japan in 1945.

IV.

Although commercial air travel began in the 1920's, it hardly succeeded until the end of the Second World War. The obstacles to transportation by air were several. In the first place, and most important, was the lack of public confidence in airplanes. This meant that few people were willing to risk their lives flying, so that the early airlines could not raise much capital to invest in the new enterprise.

The first airfreight pilot was Philip Parmelee, who flew a Model "B" Wright Flyer from Dayton, Oho to Columbus, Ohio on November 7, 1910. He flew for sixty-six minutes to deliver silk to the owner of a department store, seven years after the first flight at Kitty Hawk. That flight was a major achievement of the Wright Company.[20]

In 1910, flying was considered an entertainment rather than a serious means of transportation. The entertainment consisted of exhibitions at state fairgrounds and was conducted by such renowned flyers as Glenn Curtiss. The first air show was held in France in 1909 and the first American air show was held in Los Angeles in January of 1910.[21]

Pilots were not licensed in the early 1920s and no one certified that an airplane was safe. Indeed, the Department of Commerce launched a publicity campaign to support flying, but this had little success. Moreover, aircraft producers could hardly earn a profit, because the government no longer ordered new planes as had been done during the world war, and private airlines were so few that they could not support the aircraft industry.[22]

The problems of the aviation industry were numerous. For one, the industry suffered from such a dearth of commercial flying that there was no market for new planes. In addition, the government sold surplus planes left over from the world war cheaply, so that manufacturers could hardly compete. For example, in 1921, there were 88 airlines operating in the United States. Only 17 of these airlines still existed in 1923.[23]

It was not until 1926 when Congress passed the Air Commerce Act that the U.S. government made any effort to support the aviation industry. That law directed the Secretary of Commerce to "foster air commerce in accordance with provisions of this act." Included in this act was a provision to examine and rate airmen as well as air navigation facilities and other requirements having to do with safety, weather forecasting and other provisions, and demands and prohibitions pertaining to air travel.[24]

After the passage of the Air Commerce Act, the mileage covered by air rose from 3,369 miles in 1925 to 25,000 miles in 1929, of which 14,000 were lighted. In those three years, one thousand airports were built and 6,400 planes were licensed, as 7,500 planes were manufactured each year.[25]

The outcome of these efforts was the steady increase in air travel after the depression and the Second World War had come to an end. By 1952, American domestic airlines carried 32 million passengers for more than 16 billion miles. This meant that there was a tenfold increase of passenger miles flown since 1942, a year in which nineteen airlines operated 186 aircraft. Twelve years later, in 1954, thirty-two airlines operated 1,175 planes.[26]

Two government agencies regulated airlines starting in the 1950's:The Federal Aviation Agency and the Civil Aeronautics Board. The FAA was concerned with operations and maintenance of airways and the CAB dealt with financial affairs, mergers, and the licensing of air routes.[27]

The Civil Aeronautics Board was the product of the Civil Aeronautics Act of 1938, which created the Civil Aeronautics Authority. The name was later changed to the CAB. That agency was authorized to control routes, air fares, mergers, and all other aspects of air travel.

By 1960, U.S. airlines provided thirty billion passenger miles of domestic air service. Nevertheless, eleven lines earned only $22 million that year, and the rate of return on investment was only 2.9 per cent. This also meant that some airlines were accumulating losses and others were barely surviving. During the 1960's, airlines were converting to jet engines. This was very expensive and was therefore one of the reasons airlines were so unprofitable during that decade. Once jet travel had been accomplished, the use of air travel affected almost all business and government operations, even for Americans who had never flown. Then it was estimated that most airlines flew with only 57.5% of occupied seats. Competition between airlines, subsidies for "lean" routes, and government regulations were additional burdens which threatened profitability. As a result, a number of airlines have merged over the years, since they could not otherwise sustain themselves.[28]

As these airlines merged and some of the more prominent airline names such as TWA and Pan Am disappeared, the media, always hysterical, predicted utter disaster for airlines. Indeed, in 2005, one of the worst years for airlines, their losses exceeded $43 billion. Yet, more than 2 billion people fly each year and more than $3.2 trillion worth of cargo is transported by air each year. Foreign airlines are more profitable because they do not face as much competition as U.S. airlines and because they do not have to contend with unionized labor demands.

It has been estimated that America'sairline industry will double in size every fifteen years, so that by 2020 passenger numbers will grow from two billion to four billion and cargo will increase from 39.5 million tons to 79 million tons. Much of this increase comes from Asia but benefits American airplane production.[29]

V.

Military and commercial aircraft are augmented in the United States by about 15,000 private planes, of which the vast majority is owned by U.S. companies.

The need for flying private jets is illustrated by "Hapco Farms," which owns fruit and vegetable fields from Maine to Florida. To reach the farm communities where their business is located, the company uses a 16,000 pound Beechjet 400 that seats six passengers. The advantage of using private planes is evidently that private planes avoid large airports and have access to over ten times more airports than commercial airlines, located in smaller communities. Furthermore, private planes leave and arrive at will, so that its users are not dependent on airline schedules.

Commercial airplanes are on average twenty-five years old. Private planes are generally newer. Most important, private flying avoids standing in security lines, waiting for hours for connections, and dealing with limits on baggage. There are companies who rent or charter flights at considerable expense. For example, to charter a heavy jet round trip from Van Nuys, California to Teterboro, New Jersey costs $68,780. A "heavy" jet can seat 15 passengers. Likewise, it costs $105,000 to charter a heavy jet from New Jersey to London, England.[30]

Private airplanes play a role in politics. Beginning with primary elections, candidates for the presidency must use private air travel to reach voters in areas not accessed by using commercial airlines of large city airports. There are 5,000 smaller airports across the country which allow a candidate to make several stops on any one day during the campaign. It has been estimated that it would take half a week by commercial aircraft to visit four campaign stops which can all be reached by private plane on one day. The cost of such campaigning is significant, unless the candidate has supporters who own a plane and are able and willing to carry the candidate to his many destinations.

After the conventions, when the candidates have been nominated, the cost of air travel increases as staff and reporters are all carried on the campaign plane. Such an aircraft is usually much larger than what was used during the primaries and will cost the leading candidates for the presidency upward of $20 million. The incumbent president needs no such support, as he can and will usually use Air Force One at the taxpayers' expense.[31]

The Transportation Security Administration has introduced a number of regulations designed to prevent private aircraft from becoming a terrorist weapon or a conveyance for dangerous materials. Accordingly, operators of private planes will have to conduct fingerprint checks on pilots, cross check passengers' names against a government watch list, establish a security program, submit to regular audits, and limit the items allowed on an aircraft.[32]

One of the disadvantages of flying in private planes is the greater risk of accidents. According to the National Transportation Safety Board, private planes were involved in 22 fatalities for every million flight hours. Scheduled airliners were involved in only four accidents for every million flight hours. Commuter airlines are also more dangerous than larger commercial airlines, as their fatal accident rate ranges from 11 to 12 per million flight hours.

An example of a commuter plane crash which killed fifty people occurred on February 13, 2009, when a "Bombardier" aircraft operated by Colgan Air on behalf of Continental Airlines crashed six miles from the Buffalo, New York airport. The plane had originated at Newark Liberty International Airport. Later investigation discovered that pilot error was the principal reason for this deadly crash, although ice buildup played a part in the disaster. The National Transportation Safety Board found that the pilot and co-pilot had been inadequately trained and that the pilot had not slept in nearly 24 hours. Evidently, the plane stalled, calling for the pilot to pull the vibrating column when he should have pushed. This caused the plane to stall and fall into a house, killing one person.[33]

VI.

The need for pilot training is obvious. In 1910, the Wright brothers opened the first civilian flying school outside Montgomery, Alabama. They did so because the winter weather around their home town Dayton, Ohio would not allow them to fly for months every year. They therefore decided to take advantage of an offer by plantation owner Frank Kohn, who charged them nothing for the use of his land as an airport. There the Wrights built the first hangar and then began teaching five students how to fly. This included training in understanding wind currents and methods of balancing and operating the wing warping levers causing the airplane to bank and turn. Further, students learned how to operate the control levers which caused the airplane to ascend and to land.[34]

In time, those who learned flying from the Wright brothers became the forerunners of all who would fly thereafter. Eventually, flying lessons would begin on the ground, as electronic flight simulators were developed to reproduce in detail all the switches, dials, levers, and controls found in the cockpit of a modern airliner. The first simulator was designed by Dr. R.C. Dehmel of the Curtiss-Wright Corporation. That corporation was formed in 1929 by merging 12 airplane and engine manufacturers.[35]

The simulator cost $250,000 to build in 1948. It was preceded by a considerable amount of planning and research. The simulator allows the entire crew, i.e. the pilot, the co-pilot, the engineers, and the flight attendants, to be trained at the same time. An instructor operates the instruments and causes all kinds of difficulties, such as oil leaks, carburetor icing, and other problems. The instructor then records the pilot's response time and the corrective action taken.[36]

Because flying takes a great deal of training and involves many dangers, the Federal Aviation Administration has been authorized by Congress to certify pilots. To that end, the FAA has established a number of categories of aircraft. These categories result from the advances in technology available in the twenty-first century. They are far more complicated and difficult to learn than was true in the early days of aviation. These categories are: airplanes, rotorcraft, glider, lighter than air, powered lift, powered parachute, and weight shift control. These categories are further broken down. The airplane is divided into "single engine land," "multi engine land," "single engine sea," and "multi engine sea" classes.[37]

In addition to certifying pilots, the FAA also certifies flight engineers, although modern jets are generally operated by two persons. Likewise, flight navigators have become obsolete. The FAA also certifies flight instructors, parachute riggers, aircraft maintenance technicians, and air traffic controllers. Flight attendants are also required to earn an FAA "Certificate of Demonstrated Proficiency."[38]

Major commercial airlines train pilots as well as other personnel who service the flying public. This is a very expensive and complicated aspect of the aviation business, and is so important that it may well be said that an airline which cannot train personnel effectively cannot survive. To achieve a working training program, an airline must know the number of pilots needed for its flight schedule. Evidently, airlines must employ flight instructors, develop training curricula, and determine how much work can be assigned to each pilot. Pilots are generally divided into four groups. These are pilots who advance without additional training, pilots who advance with further training, pilots who move to a new base but continue to fly the same type of aircraft such as DC10 or 737, and pilots who move but retain the same status such as captain, first officer or second officer.[39]

Despite the extensive training afforded every pilot, Rory Kay, co-chair of the FAA committee that examines pilot training, said in 2011 that airlines suffer from "automatic addiction."Kay claimed that "we are forgetting how to fly" because almost all flying tasks are handled by computers.[40]

The training of pilots in the United States is achieved by both civilian flight schools and by the military. The difference between the two sources of our pilots is immense. The air force trains pilots at the Air Force Academy in Colorado Springs, Colorado, and the navy trains pilots at the School of Naval Aviation at Pensacola, Florida.

The curriculum at the U.S. Air Force Academy includes several possible "majors" in such subjects as chemistry, physics, and mathematics, as well as the study of foreign languages and the behavioral sciences. In addition, the Department of Aeronautics teaches numerous courses such as "Fundamentals of Aeronautics," "Aero-Thermodynamics," "Aero Engineering," which is extensive, and deals with a variety of aspects of engineering followed by extensive training in "Airmanship" which deals with basic piloting, advanced piloting, a powered flight program, soaring and parachuting, and more. In fact, anyone who has graduated from the Air Force Academy or the School of Naval Aviation has learned so much more about flight than graduates of private pilot training schools that a comparison is impossible. Therefore, graduates of the Air Force Academy are preferred candidates for jobs as commercial pilots. Although graduates of two year junior colleges have been appointed to fly commercial airplanes, the majority are college graduates for a job which pays an average of $103,000 a year.[41]

The training of air traffic controllers is as important as the training of those belonging to a flight crew. Air traffic controllers have a strong union and regard themselves as professionals. The union is called the National Air Traffic Controllers Association. In 1981, shortly after Ronald Reagan became president, 13,000 members of NATCA's predecessor, the Professional Air Traffic Controller's Organization (PATCO), went on strike when no agreement concerning several issues could be reached with the Federal Aviation Agency. This caused President Reagan to fire 11,350 controllers, or almost 70% of the work force. Those who did not strike were retained and supported the system until technology and the training of new controllers returned the system to its erstwhile strength. The 23,400 controllers earn an average of $117,400.[42]

There are about 90,000 flight attendants working in commercial aviation. Their earnings are approximately $37,740 a year. Training for flight attendants range between two and eight weeks. A few airlines pay a salary during flight attendants' training, although no airline considers a trainee an employee. Training for these jobs is also undertaken by a host of private "cabin crew" training schools claiming that they are a sure route to a flight attendant's job. In fact, much of what these schools do is superfluous, because the airlines will train their own employees because every airline operates its own aircraft, so that training and work conditions differ depending on the type of place in use. The training schools do however prepare the candidates for the employment interviews which are mandatory for all applicants.

The job of flight attendant is primarily to insure the safety of the passengers and secondarily to insure their comfort. This means that flight attendants must be alert and physically fit to deal with emergencies. Normally, flight attendants distribute snacks and, on very long flights, meals. They also help disabled passengers, deal with small children, and administer first aid if needed.[43]

VII.

It has been estimated that seventy-five percent of Americans have flown in an airplane in their lifetime and that therefore twenty-five percent have never flown. According to the Bureau of Transportation Statistics, about 100 million Americans fly in the course of a year. A far greater number would include foreigners who fly on American planes. A number of the 100 million passengers fly more than once each year, so that the BTS estimates that 637 million passenger flights occurred in 2011, an increase of 1.1% since 2010.[44]

Evidently, the experience of flying has transformed transportation in the course of a century. Throughout all of the history of mankind, flying was impossible, so that only a small minority of humans have ever participated in this relatively new experience. Since the day when the Wright brothers first flew a heavier-than-air powered plane in 1903, vast advances were made in aerodynamics, propulsion, avionics, navigation, and the material sciences. These advances allowed commercial aircraft to fly longer, faster, and higher than ever before, leading to the compression of time and distance.[45]

The ever greater development of air travel is not only a means of moving people between places, but is also a means of creating new social practices at airports and in aircraft. There are in airports globally franchised retail chain stores.Airport lounges serving arrivals and departures have become more than just places to wait. Instead, the airport has a different meaning for those who use it. This means that a flight crew, passengers, baggage handlers, controllers, and others all have different experiences at airports or in flight. Passengers certainly do not have the same experiences than is true of pilots, who contend with pitching, rolling, yawing (sliding from side to side), and other sensations not felt by the passengers.

Prior to 1930, air travel was most unusual, so that people read the published accounts of air travelers so as to learn about the sensations and conditions of flight. The experience of flying changed rapidly year by year as aircraft changed. The flying experience is also dependent on the route flown and the condition of the passengers. Some passengers became nauseated by the motion of the 'plane and others were and are "white knuckle" flyers in that they experience fear and even panic before and during a flight. This means that for some passengers a flight must be endured rather than enjoyed. In fact, in the early days of flying, cabins were cold and not pressurized, so that passengers had to wrap themselves into sweaters and coats to tolerate the discomfort. Even now, passengers spend most of their time on a plane seated, so that on a long flight this becomes most uncomfortable. Sitting for hours in the same position leads to cramps in the legs, so that some passengers exercise in the aisle of the plane.[46]

In the early days of flight, passengers were fearful of crashing, a fear accelerated by turbulence, leading planes to suddenly lose altitude or slide from side to side. Of course, these fears were offset by those who viewed the flying experience in a positive light as a means of observing the earth in an unaccustomed way. The sensation of seeing things from a great height led some passengers to write about these experiences for those who had not yet flown. Just as the steam engine created a new world for those who traveled across the country in three days, the airplane created a new world for all who experienced its novelty in the twentieth century.[47]

VIII.

In 1976, Senator Edward Kennedy of Massachusetts held hearings on the federal government's regulation of the airlines. The principal focus of these hearings was the cost of flying, which Kennedy held to be far too high because the Civil Aeronautics Board had prohibited airline price competition. The evidence indicated that wherever price competition was allowed, fares were much lower. An example was the cost of flying from San Francisco, California, to Los Angeles, a distance comparable to flying from New York to Washington, D.C. The California route cost a good deal less than the route on the east coast because the San Francisco to Los Angeles route was not regulated. President

Gerald Ford was willing to consider the deregulation of the airlines, but was succeeded by Jimmy Carter, who signed the Airline Deregulation Act in 1978.[48]

The Civil Aeronautics Board is the successor of the Civil Aeronautics Authority, which came into being in 1938 when Congress passed the Civil Aeronautics Act, which was the first legislation dealing with air transportation.[49]

As a consequence of deregulation, airlines have been able to charge far higher fares than the CAB ever permitted. This is particularly true of travel in monopoly markets, in which one airline dominates the market. In those situations, airfares are eighteen to twenty-seven percent higher than on routes which are subject to competition. In time, a number of airlines have ceased to be in business as they were driven out by the competition. Deregulation allowed airlines to cease operations in smaller communities and service only larger populations.[50]

IX.

The most recent experiences of airline passengers have less to do with viewing the world from a great height than with the annoyance which security issues impose on passengers in the United States and around the world. As a consequence of the attack on the World Trade Center in New York City on September 11, 2001, considerable changes in airline security have been instituted by the United States government. These include detailed examinations of passengers, their possessions, and the baggage they deliver to the cargo areas of planes they intend to board.

Included in the precautions taken at airports in an effort to prevent terrorism are so-called "pat down" procedures, which annoy women even more than they annoy men. According to complaints by passengers concerning these methods, they are so invasive that they resemble "groping" by the employees of the Transportation Security Administration. Writing in the New York Times, a lady lawyer described a humiliating experience at the Tampa, Florida Airport as TSA screeners inspected a lady's underwear and even touched her breasts.[51]

Many of the complaints concerning the TSA are rooted in the American distrust of government and the unwillingness of citizens to take orders from bureaucrats. An example is a most recent rule that the name on a photo ID must be exactly the same as the name on a boarding pass. This would mean that someone with a middle initial must use that initial also on the boarding pass. Likewise, passengers must remove their shoes at inspection areas and unload all metal items, including loose change. Absurd carry-on restrictions have also been put in place, including prohibiting liquids and gels. Most objectionable to passengers are full body screening machines which show an image of the naked body. These machines make it possible to move through the maze of security measures much faster than without their use. However, many passengers resent the invasion of privacy which these machines represent.[52]

There are numerous stories told by passengers about "pat downs," missed flights, delays, and harassment at airports due to the inspection procedures.

Indeed, some "experts" claim that these procedures are not needed and that only the bolstering and strengthening of the door leading to the cockpit or the pilot's cabin is sufficient to secure the airplane from a terrorist's mid-flight attack. Others favor that pilots carry guns or that ethnic "profiling" be used before passengers are allowed to board. In sum, the security measures taken after the September 11, 2001 attack are unpopular but evidently needed, as there have been numerous threats against American airlines for eleven years.

Of course, the attacks on the World Trade Center were not the only attacks on the American airline industry. In the 1960's and the 1970's, numerous hijackings of airplanes diverted to Cuba led to an agreement between the United States and Cuba in 1973, calling for the hijackers to either be returned to the United States or be imprisoned in Cuba.In 1983, five airplanes were diverted to Cuba, and a number of others continued until the Cuban authorities jailed the hijackers for ten to twenty years.[53]

The first airplane forced to land in Cuba had left Marathon, Florida on its way to Key West, Florida when a man with a knife and a gun forcedthe plane to fly to Havana. On July 31, 1961, Bruce Britt attempted to force a Pacific Airlines flight at Chico Airport in California to fly to Arkansas. The attempt failed, but Britt shot two employees of the airline, blinding one for life.[54]

Cuba is home to a number of Americans sought by the F.B.I. who came there in the 1970's after hijacking a TWA plane at gunpoint. One of these is Charles Hill, who killed a New Mexico trooper, Robert Rosenbloom, who had stopped him on a remote highway. Hill claims that Rosenbloom was a racist and that he killed Rosenbloom in self-defense. This murder and hijacking occurred before the Cuban government gave hijackers stiff sentences under an agreement with the United States.[55]

Numerous additional hijackings and attacks on airplanes marked the 1980's, topped by the destruction of a Pan Am Boeing 747 which exploded over Lockerbie, Scotland on December 21, 1988, killing 259 passengers and crew and also killing 15 people on the ground. Thirty-eight of the passengers were students at the Syracuse (New York) University. The plane was flying from London to New York when it exploded.[56]

In 2000, a Scottish court convicted Abdelbaset Ali Mohmed Al Megrahi, a Libyan "intelligence agent," of planting a suitcase holding a bomb on the plane. It was alleged that Megrahi acted on orders of the former Libyan dictator Ghaddafi.In 2009, Megrahi was released on the grounds that he was terminally ill. He returned to Libya amidst claims that he was "framed" and that others were "behind" the bombing of the Pan Am plane.[57]

The use of airplanes as weapons of mass murder had not been known until September 11, 2001. At 8:44 a.m. on that day, American Airlines Flight 11, a Boeing 767 airplane with 93 people aboard, crashed into the World Trade Center upper north tower in New York City. The plane had come from Boston and was en route to Los Angeles. During that short flight it was hijacked by Arab terrorists and flown into the office building, with the result that black smoke poured from the gaping hole in the flaming tower. Nineteen minutes later, at 9:03 a.m., United Airlines flight 175, a Boeing 757 with 65 people aboard, struck the south tower of the World Trade Center. The plane came from Washington, D.C. en route to Los Angeles. Because the first attack had brought television coverage to the area, the second attack was viewed by millions in horror.

Forty minutes after the second attack on the World Trade Center. American Airlines Flight 77, a Boeing 757 with 64 people aboard, crashed into the south ring of the Pentagon building in Arlington County, Virginia.

Even as the south tower of the World Trade Center collapsed , raining debris and a cloud of dust and rubble all over the area below, United Airlines flight 93, with 92 people aboard, crashed into the forest in Somerset County, Pennsylvania, killing all passengers and crew members.[58]

The number of people killed in these attacks was approximately 3,000. In addition to the people killed in the airplanes, 125 people were killed in the Pentagon attack. Inside the World Trade Center 2,925 died as the towers collapsed, and some jumped to their deaths because of fires in the buildings. Those who died in these attacks came form seventy-seven countries and every state in the union. Small children were killed, as were the old and the middle aged. More men than women died in this inferno because more men were doing business in the trade center.[59]

X.

Air travel was totally changed by the events of September 11, 2001. Furthermore, air travel in the early 21st century has also been radically altered by economic conditions affecting the airlines. While at one time passengers were given substantial meals even on short flights, food has been restricted to a few peanuts and a drink unless the trip involves several hours. Moreover, many flights have been canceled, forcing travelers to connect at the most crowded and largest airports, even as ticket prices have skyrocketed due to higher oil prices, which reached $ 100 a barrel in the spring of 2012, with the expectation that the average will reach $115 a barrel by the fall of that year. A lack of competition is another factor promoting increased costs for air travelers. The economic downturn of the Obama years has led to a decline in business trips by 22 percent between 2008 and 2013. Nevertheless, the cost of travel has increased by 3.6% in that same period.

Those who forecast economic conditions believe that spending by American business will increase by $260 billion in 2012-2013 despite the assumption that business trips will be reduced from 445 million in 2011 to 438 million in 2013. Inflation has led to an increase in spending from an average of $422 in 2000 to $564 in 2011.

Increased spending for air travel was mandated by an average fare rise of 10% to $254 in 2011, rising to $250 by February of 2012. Hotel rooms also rose three percent during those months.[60]

Air travel is a business. Therefore, the future of air travel can only be assured as long as the operation of airlines is profitable. Accordingly, it is significant that the airline industry globally lost $9.3 billion in 2010. Airlines recorded an increase in airline profits of 16% in 2011 to $560 billion globally. North American airlines earned $3.5 billion in 2011. In 2012 the airline industry appears to be in serious difficulty. This is true because the passenger market has not increased in the first quarter of 2012 while the freight business is shrinking. In fact, travel in the United States fell by 0.3% in 2011.Meanwhile, the high cost of jet fuel increased, so that these fuel costs will add $15 billion to the expenses of running the airlines. The gross domestic product usually predicts the solvency of airlines. In 2011, the GDP rose only 2.4%, and no more is expected in 2012. Finally, airlines, like all American businesses, are so burdened by taxes and endless regulations that government is slowly throttling the airline business, which cannot survive without a profit.[61]

XI.

Airplanes and air travel are the subject of a good deal of entertainment, as is visible by considering the numerous movies concerning aviation. Many of these films are war movies involving the exploits of military hero-aviators while others are disaster films seeking to exploit public enjoyment of last minute rescues and heroic crews, if not passengers. Included also are some biographies such as "The Aviator," which is a story about Howard Hughes, a multimillionaire who invested heavily in airplane production. He bought a majority interest in TWA, or Trans World Airlines, and subsequently flew faster than anyone before him. He also flew around the world in four days, shattering the previous record by four days. Hughes also invested in building a large 'flying boat" which he called "Spruce Goose," but which crashed on a test flight, leaving Hughes seriously wounded.

Another of the more than sixty movies dealing with airplanes is "Air Force One," which is so ridiculous that it fits the most common form of Hollywood entertainment based on mindless violence and a plot of utter nonsense.

Three movies called "Airport" were produced in the 1970's and are listed as "drama-disaster" movies, as are "Executive Decision," "Fight Club," and "Five Came Back."

"Pearl Harbor," released in 2001, deals with the Japanese attack on December 7, 1941. That historical tragedy serves as the background for the usual "romantic triangle" found in so many Hollywood productions. The airplane angle is the air attack on the part of the Japanese. A similar movie called "Tora! Tora! Tora!" made in 1970, also depicts the attack on Pearl Harbor from the Japanese and American view. The movie is fairly accurate in depicting the events of that day.

Also of historical interest is "The Spirit of St. Louis," which dramatizes the flight of Charles Lindbergh from Roosevelt Field on Long Island, New York to Paris in 1927. Flying for 33 hours in a monoplane and without a parachute,

Lindbergh achieved a historic goal of flying across the Atlantic for the first time in aviation history. This movie has some merit in that it recites an event which made the now commonplace crossing of the Atlantic Ocean possible.

A recent movie concerning air travel is "Up in the Air," released in 2009. The principal message concerns the horde of traveling salesmen and saleswomen who spend nearly all their time flying around the country while losing all connection to family and friends as they move from airport to airport, from city to city. The movie includes the additional message concerning the impersonal cruelty of the corporate world as the protagonist in the movie, played by George Clooney, travels about with a female associate, firing long time employees from jobs they have held for years in the expectation of retiring from the company that has used them for so long. The movie shows the desperation of those "laid off," and includes a suicide among those dismissed. This movie is indeed a commentary on American society in the 21st century, largely controlled by technical devices and lacking human interaction and relationships.[62]

Summary

Ancient legends as told by Babylonians, Hindus, Greeks, Chinese, and others reveals that mankind had always sought to fly. Yet no flying became possible until the balloon was developed in France and the glider in Germany. Yet, real heavier-than-air flying became possible only after the Wright brothers invented the powered airplane in 1903. Consequently, the early airplanes were used in the First World War, and thereafter to carry mail. Passenger flying did not become popular until the 1930's and grew exponentially after the Second World War.

As air travel became more prominent and greater numbers of people flew, government began to regulate the airplane industry more and more. Consequently the income of airlines became dependent on the gross national product as well as the regulations imposed on them. Nevertheless, flying has become a necessity in American business, as shown in such movies as "Up in the Air."Other movies concerning air travel are also a form of entertainment among American audiences.

Notes

1. James P. Harrison, *Mastering the Sky: A History of Aviation From Ancient Times to the Present,* (New York: Saperdon, 1996), pp. 2, 3, 31.

2. Ibid, 14.

3. J. Gordon Leishman, *Principles of Helicopter Aerodynamics,* (New York: Cambridge University Press, 2006), pp. xx.

4. Harrison, *Mastering,* 24-25.

5. C. C. Gillespie, *The Montgolfier Brothers and the Invention of Aviation, 1783-1784,* (Princeton, NJ: Princeton University Press, 1983), pp. 178-179 and 183-185.

6. Viktor Harsch, "Lillienthal's Fatal Glider Crash, in 1898," *Aviation, Space and Environmental Medicine,* (October 1, 2008), pp. 993-994.

7. Fred C. Kelly, *The Wright Brothers,* (New York: Harcourt, Brace and Co., 1945).

8. George F. McLaughlin, "The Story of the Airplane Engine," *The Annals of the American Academy of Political and Social Sciences,* vol. 131, (May 1927), p. 35.

9. Roger E. Bilstein, "The First Air War 1914-1918," *The Journal of American History,* vol. 78, no.4, (March 1992), p. 1483.

10. Walter LaFeber, "A Note on the 'Mercantalistic Imperialism' of Alfred Thayer Mahan," *The Mississippi Valley Historical Review,* vol. 48, no.4, (March 1962), p. 674-685.

11. Edward Goldstein, "Wings of Gold: One Hundred Years of U.S. Navy Air Power," *Aerospace America,* (September 2011), p. 22.

12. Walter Musciamo, "German Condor Legion's Tactical Air Power, *Aviation History,*(September 2004), p. 2.

13. Mark Harrison, *The Economics of World War II,* (Cambridge, UK, 1998), p. 85.

14. Stephen B. Sweeney, "Some Economic Aspects of Aircraft Transportation," *The Annals of the American Academy,* vol. 131, (May 1927), p. 162.

15. William A. Steiger, "Lindbergh Flies Air Mail from Springfield," *Journal of the Illinois State Historical Society,* vol.47, no. 2, (Summer 1954), p. 140.

16. Ibid., p. 147.

17. Raymond Clapper, "Air Mail Action Viewed As End of Subsidies for Favored Few," *Washington Post,* (February 11, 1934), p. 2.

18. No author, "Army Ill-equipped For Mail Service, Officer Predicted," *San Diego Union,* (March 18, 1934), p. 5.

19. James A. Mackay, *Airmails 1870-1970,* (London: B. T. Batesford, 1971), p. 216.

20. Alice Murphy and Peter Murphy, "Philip Parmalee: the World's First Commercial Pilot," *Michigan History Magazine,* vol. 95, no. 1, (January-February, 2011), p. 22.

21. Charles H. Gibbs-Smith, *Aviaton: An Historical Survey from Its Origins To the End of World War II,* (London: Science Museum, 1970), pp. 152-158.

22. Elisabeth S. Freudenthal, *The Aviation Business from Kitty Hawk to Wall Street,* (New York: The Vanguard Press, 1940), p.60.

23. "Aircraft Production in the United States, *S. Rept. 555, Congressional Record,* 65th Congress, 2nd Session, 9239-34.

24. *The Statutes At Large of the United States of America,*vol. XLIV, Part 2, Sixty-ninth Congress, 1925-1927, Session I, Chapter 344.

25. Freudenthal, *Aviation Business,* p. 77.

26. William J. Megonnell and Howard W. Chapman, "Sanitation of Domestic Airlines," *Public Health Reports,* vol. 71, no. 4, (April 1956), p. 360.

27. Samuel A. Richmond, "Why Are The Airlines In Trouble?" *Challenge,* vol. 10, no 8. (May 1921), p. 6-7.

28. Ibid., p. 8.

29. Giovanni Bisignanni, "Airlines," *Foreign Policy,* (January/February 2006), p. 22.

30. Jacob Leibenluft, "Six Thousand Gallons of Regular, Please," *Slate Magazine,* (April 28, 2008), p. 1.

31. Jo Sharkey, "Private Jets Warming Up for Presidential Campaign," *The New York Times,* (April , 5 2011), p. 6.

32. Christine Negroni, "Business Jet Owners Say Proposed Rules Won't Improve Security," *The New York Times,* (January 11, 2009):section LI, p. 5.

33. Dale Anderson and Phil Fairbanks, "49 killed as Plane Crashed into Home in Clarence Center," *The Buffalo News,* (February 13, 2009), p. 1.

34. Jerome Ennels, ""The Wright Stuff: Pilot Training at America's Frst Civilian Flying School," *Air Power History,* vol. 49, no. 4, (Winter 2002), p. 22-31.

35. Curtiss-Wright, "The Spirit of Innovation," http://www.curtiss-wright.com/history.asp.

36. No author, "Flight Training on Ground," *Science News Letter,* vol. 53, no. 21, (May 22, 1948), p. 325.

37. U.S. Department of Transportation, Federal Aviation Administration, CFR Part 61, "Certificaton: Pilots, Flight Instructors and Ground Instructors," Sib-part A General Section 61.5.

38. Ibid.

39. Gang Yu et. al. "Optimizing Pilot Planning and Training for Continental Airlines," *Interfaces,* vol. 34, no. 4, (July-August 2004), p. 253-264.

40. Jeff Van West, "It's About Engagement, " *IFR.* Vol. 27, no. 11, (November 2011), p. 2.

41. http://af.mil.

42. Joseph McCartin, "Ronald Reagan, the Air Traffic Controllers Association and the Strike that Changed America," *The New York Times,* (August 3, 2011), Sec. A, p. 25.

43. http://www.flightattendantcabincrewtraining.com.

44. U.S. Bureau of Transportation, Bureau of Transportation, Research and Innovative Technology Administration, "Trans Stats," (Washington, DC, 2011), p. 1.

45. John Bowen, "Network Changes, Deregulation, and Access in the Global Airline Industry," *Economic Geography,* vol. 74, no. 4, (2002), p. 425-440.

46. Gordon Pirie, "Cultural Crossings," *Journal of Transportation History,* vol. 29, no.1 (2008), pp. 1-4.

47. Lucy Budd, "The View from the Air," In: P. Vanini, *The Culture of Alternative Mobilities,* (Ashgate, Farham, 2010), pp. 71-90.

48. Steven Breyer, "Anti-deregulation Revisited," *Business Week,* (January 20, 2011).

49. No author, "Civil Aeronautics Board Policy: An Evaluation," *The Yale Law Journal,* vol. 57, no. 6, (April 1948), pp. 1055-1956.

50. B. Parolin and A. Harrington, "Changing Long Distance Passenger Markets in a Deregulated Environment," *Transportation Research Record,* vol. 1341, (1992), pp. 28-36.

51. Joe Sharkey, "When a Pat-Down Seems Like Groping," *The New York Times,* (November 2, 2004), C, p. 9.

52. Christian Negroni, "What to Expect as Airport Security Rules are Tightened," *The New York Times,* (January 2, 2010):Sec.A, p. 10.

53. No author, "Hijackers To Cuba Given Stiff Terms," *The New York Times,* (July 7, 1983):Section B, p. 7.

54. Greg Welter, "First Sky Jacking Attempt Was In Chico, 45 Years Ago," *Chicoer,* (July 31, 2006), p. 1.

55. Serge F. Kovaleski, "Havana is Haven for Fugitive '70s Hijacker," *Washington Post,* (August 31, 1999), p. A07.

56. Edward Cody, "Pan Am Plane Crashes in Scotland, Killing 270," *Washington Post,* (December 22, 1988), p. AO1.

57. Editorial, "Megrahi Inquisition," Daily Mail, (December 10, 2010), p. 1.

58. No author, "A Day in the Life of America," *The New York Post,* (September 12, 2001), p. 18.

59. No author, "Deaths in World Trade Center Terrorist Attacks," *Morbidity and Mortality Weekly Report,* (Atlanta, GA: Centers for Disease Control and Prevention, September 11, 2002), pp. 1-4.

60. Jad Mouawad, "Trapped in the Middle Seat," *The New York Times,* (May 3, 2012): Section F, p. 1.

61. Susan J. Aluise, "Reasons Airline Earnings Will Tank in 2012," *Investment Place Media,* (October 10, 2011), p. 1.

62. http://www.ranker.com/list/airplanes-and-airports-movies-and-films/reference.

Chapter Ten

The Computer, the Internet, and the World Wide Web

I.

It is not unreasonable to view the ten fingers on the human hand as a computer which allows a child to learn how to count to ten. In an effort to help with counting, a number of Asian peoples developed a device called an "abacus." That word came from the Hebrew and the Greek and finally entered Latin. In Hebrew the word means "dust" and refers to the dust or sand covering a surface used to draw geometric diagrams. Later the abacus became a frame holding rods on which a number of beads move up and down. Each rod is designated as a denomination such as tens, hundreds, or thousands in the decimal system. Each bead represents a digit allowing the user to add, subtract, multiply, and divide. Because the abacus has no memory, it is limited in its applications, unlike the digital computers used today. Nevertheless it was used for centuries in almost all Asian civilizations and is still in use today.[1]

Another "computer" is the quadrant. A quadrant, as the name implies, is one quarter of a circle, which can measure angles up to 90 degrees. It has been used to measure astronomical distances as well as geometric distances, and is used in navigation as well as surveying. In 1623, Edmund Gunter invented a quadrant which can be used to determine time. Trigonometric problems can also be solved by using a quadrant. It was also used to make solar observations, and has been framed for that purpose in several places, including Beijing in China.[2]

Gerbert d'Aurillac (955-1003) taught mathematics in Rheims, France, and introduced the Hindu numbers to Europe. He also compiled a list of rules for the use of the abacus in which he explained how to multiply, divide, add, and subtract. He also invented the astrolabe, which was used to determine the altitude of the sun and other celestial bodies. Gerbert became pope Sylvester II in 999.[3]

In the thirteenth century, Roger Bacon (1214-1294) and Albertus Magnus (1206-1280) built so called "speaking heads," and Leonardo da Vinci (1452-1519) drew a pedometer in 1519. A pedometer measures the distance traveled on foot by counting the number of steps taken.

William Schickard (1592-1635) sent several drawings of a calculator clock to Johannes Kepler between 1623 and 1624. These letters preceded the invention of Blaise Pascal's calculator by twenty years, although Schickard's letters were lost for three hundred years and had no impact on the later development of the calculating machines and the eventual development of the computer.

Several failed attempts were made to construct a calculating clock until Giovanni Poleni succeeded in 1709 to construct a two motion calculating clock.[4]

Although there has been some controversy in favor of Schickard, there can be little doubt that it was Blaise Pascal who invented the first truly working mechanical calculator in 1642. He perfected his machine in 1645 and built a number of these machines thereafter. This led to the production of similar machines all over Europe, including the machine built by Gottfried Leibniz (1646-1716). Leibniz had invented the calculus independently of Newton and was known as a polymath. Leibniz improved on the "Pascaline" and Thomas de Colmar based his arithmometer in 1820 on both the work of Pascal and Leibniz. Susequently Dorr Felt (1862-1930) and David Roth continued improvements through the nineteenth century.[5]

In 1884, William Hollerith, a German immigrant to the U.S., invented a device for storing data on cards through a series of holes, each representing a different piece of information, such as age, education, religion, and location. The cards were sorted by machines to produce cross tabulated data. These punch cards became the means by which International Business Machines succeeded in developing worldwide expansion, including a branch in Germany. The supply of these cards was tightly controlled by I.B.M., whose chairman & CEO, Thomas J. Watson, visited Adolf Hitler, the German dictator, in 1937, and sold him 90 million punch cards designed to record the existence and location of every Jew in Germany and the territories conquered by the German armies. This allowed the Nazi killing machine to transport millions of Jews to the gas chambers. Evidently, every concentration murder camp had a Hollerith machine. I.B.M. made a great deal of money from this bloody business, and Watson was awarded a medal by Hitler for his service to the German cause.[6]

It would be unreasonable to list here everyone who contributed to the development of the electronic computers which came into use at the end of the twentieth century. Suffice it to mention Joseph Marie Jacquard, who in 1801 invented a textile loom which used a series of punched cards allowing the loom to weave patterns automatically. When in 1837 Charles Babbage invented the first mechanical programmable computer, he created the foundation of computing for increasingly sophisticated analog computers in the twentieth century. Then, in 1936, Alan Turing became "the father of computing science" by describing what he termed a Logical Computing Machine with an unlimited memory capacity, which later led to the electronic digital computer. Time magazine named Turing one of the most influential people of the 20th century when that magazine wrote: "The fact remains that everyone who taps at a keyboard ... or a word processing program is working on an incarnation of a Turing machine."[7]

There followed the Atanasoff–Berry computer, which was the first electronic digital computer, but which was not programmable. Konrad Zuse and Tommy Flowers also contributed to the electronic computer, but it was George Stibitz who is credited with building the first modern digital computer. So many men participated in the eventual development of the present computer that no accurate definitive decision can be made as to who "invented" the computer.[8]

There is, however, no doubt that the most dramatic invention elevating computer use to its present condition was the invention of the World Wide Web by Timothy Berners-Lee and Robert Cailliau in 1989. Both men were working at the CERN (Conseil Europeen pour la Recherche Nucleaire, or European Laboratory for Particle Physics) in Switzerland. The "World Wide Web" is an internet based collection of documents and files connected to one another by hyperlinks in an immense electronic library from which anyone can retrieve information on any subject instantly.[9]

The success of the computer is best illustrated by its wide acceptance. It has been estimated by the U.S. Department of Labor, Bureau of Labor Statistics, that about 77 million Americans use a computer every day. That use is unevenly distributed among various occupations. Eighty percent of managers and professionals are most likely to use computers on the job and 67 percent of managers and professionals use the internet extensively. Government employees have an above average likelihood of using computers, as the proportion of government employees using a computer is approximately 69 percent. Over two thirds of sales and office workers use computers and the internet. By contrast, only about twenty-five percent of employees in manufacturing, transportation, and menial work use a computer or the internet. In fact, some occupations never require the use of the computer.

The demographics of computer use show that the youngest and the oldest workers were the least likely to use a computer on the job. That means that among those less than seventeen years of age as well as those over sixty-five, only one third use a computer and only 20% use the internet.

Women are more likely to use computers than men. This is true because nearly three fourths of employed women are in management, sales, office occupations, or the professions. Men are less likely to be working with computers, because two fifths of all men are employed in construction, factory work, maintenance, or transportation.

Americans with a college education are more likely to use the computer and access the internet than those with less education. In fact, 87% of college educated employees use the computer. At the other end of the education scale, we find that only 17% of those with less than a high school education use computers at work. Likewise, educated Americans use the computer to search for jobs. These jobs include computer programmers, who earn an average of $71,000 annually. Over 363,000 Americans hold such jobs. The 120,000 database administrators earn an average of $73,000 annually and the 70,000 computer hardware engineers earn $100,000. Then there are 347,000 network administrators with annual earnings of about $69,000, 308,000 computer information systems managers at $116,000 annually; about 28,000 computer research scientists who earn $101,000, and a large number of computer support specialists, estimated at 565,000, who earn about $47,000 each year. All but the support specialists must be college graduates. Computer research scientists usually hold a master or doctor degree.[10]

Because computers are used for diverse applications, a number of computer programming languages have been developed.There are about 200 such languages now in common use. The programming languages are used to write computer programs, which allow the computer to control essentially any type of electronic device, such as disk drives, printers, or robots. Programming languages differ from natural languages in that natural languages lead to communication between people while programming languages facilitate communication of algorithms, i.e. sequential instructions, to machines.[11]

II.

A vast literature has developed over the past thirty years praising the advantages and successes gained from the use of the computer by millions of Americans and people around the world. These advantages and successes cannot be denied, although the computer has also created a good deal of discomfort and pain which its proponents like to overlook. The principal downside of computer use has been the ever increasing impersonality of human relations, in that the machine creates more and more social distance between people because, for example, the computer age has produced robots or androids.

The word "robot" was invented by the Czech playwright Karel Capek and refers to a machine that looks like a man and does all kinds of work. More recently, the term robot has also been applied to machines that are computers programmed to deliver food to hospital patients or clean the kitchen floor. An "android," as the word indicates, is an artificial man (or woman) which is also a computer, "programmed" to eliminate loneliness and give us companionship. For example, there is an "android" called Roxxxy which delivers sex for $3,000. This machine was invented by an engineer and seeks to alleviate the dreadful loneliness that drives some to buy such a "toy."Evidently, so many people are having major difficulty dealing with others that this type of gadget has a real commercial future.[12]

Social robots, or computerized toys, are treated like people by many users. This is by no means new. Children have viewed inanimate dolls as "real people" for centuries, and almost all dog owners treat dogs as members of the family, attributing all kinds of human qualities to them.

Another aspect of computer provoked human relations is the permanence of computer stored online information which is never erased. Millions of people have recorded all kinds of information about themselves on Facebook and other such social networking services. That information is often detrimental to the "face" involved but remains permanently stored in the World Wide Web. Many an adolescent has recorded there all kinds of misconduct which is then visible years later when potential employers and possible romantic interests view this record to the detriment of the "face." This means that Facebook users have an online identity which may well be entirely different from their personae.[13]

As the use of texting and email increases, we no longer talk on the telephone but send electronic messages which are devoid of any emotion or feeling and limit expressions to whatever is available on the latest technological device. In fact, technology imitates interaction even when we are absent. Many users of these devices believe that they have no time to talk to anyone and therefore confine themselves to emails and other kinds of electronic communication. The numbers of those using electronic communication is increasing steadily, so that for some, personal relations become obscure and out of date. Surely there are already a good number of people who no longer have any interactive skills and who do not know how to approach other people. A good example of the manner in which electronic communications has progressed is the invention of Skype, which is a software application allowing subscribers to make free phone calls to 75 million people around the world. It also has other applications.[14]

Yet, despite the oft repeated complaints about the impersonality of all electronics driven devices, there are many who claim that their social life has improved because they have met more people on the internet than was possible for them without such help.

One of the most popular uses of the computer is the playing of video games. In fact, millions of people play video games every day without adverse effects. However, there are a considerable number of video game players whom we can call video addicts and who jeopardize their school and/or work commitments. Those who are so addicted do not use chemicals but are "behaviorally" addicted in that they need to play these games to the exclusion of all their normal responsibilities. "Behavioral" addiction consists of ever greater tension alleviated by engaging in the behavior, only to gain temporary relief. Other behaviors which may be called "addictions" are overworking, gambling, and other obsessive-compulsive disorders.[15]

It is of course possible that the non-toxic addictions are symptoms of other, unrecognized problems with which the computer "addict" is confronted. It may therefore be better to label excessive use of video games and other computer uses as compulsive rather than addictive behavior. Evidently, people who spend vast amounts of time playing video games seek to deal with depression and other moods and use these games as a means of altering their emotional state. Some of those who are compulsive video game players spend as much as 80 hours a week at this activity. This is most unusual. However, spending 30 hours a week at such activities is by no means exceptional, although the average player spends only 22 hours per week at such games.[16]

Playing video games makes the player dependent on the computer which governs it. This means that the player must respond to the predictable moves of the software. The player must also be concentrated on the game, with emphasis on eye-hand coordination. The reward of winning is mostly peer group attention, although those who play alone are rewarded by a sense of satisfaction.[17]

There has been considerable controversy concerning the effect of violent video games on violent conduct by those who have engaged in that pastime. The arguments against violent video games began as soon as the first such games became available in the 1970's. The first of these violent games was "Death Race," which appeared to allow the player to use a car to run down and kill pedestrians. Other such games followed, including "Wolfenstein," "Mortal Kombat," and "Street Fighter," all of which included one-on-one violence. Those opposed to these video games claimed that young people, particularly boys, would become violent after dealing with these computerized games. Over the years, this argument has led researchers to make several objective studies of the possible correlation between violent behavior and a personal history of playing violent video games. The sum of these studies tells us that whereas there is a positive relationship between using video games and feeling aggressive, the games themselves had little effect on the users. As is true of violent television scenes, there is no evidence that viewing violence causes violent conduct, except that those already prone to violence may be aroused to commit a violent act. Such arousal does not lead to more violent behavior among users of violent video games any more than to violent behavior by non-users. Indeed, some who have learned violence in the family or are consistently violent may very well be influenced to engage in a type of violence seen in video games. However, non-violent persons are not converted to violence by reason of any video game.[18]

III.

Another popular use of the computer is computer "dating," which refers to the advertised effort to meet the opposite sex and "jump start your love life" without wasting time. The computer driven dating services which have become near universal in their appeal to single adults are called "online communities," although the members don't know each other. These dating services advertise that their methods of connecting single people are far better than the bar scene and that they increase the member's social life. Further, it is claimed by online dating services that they cost a lot less than a so called night on the town, and that the number of people one can meet in these circumstances is far greater than is possible without these computer driven methods. Unlike the bar scene, computer driven dating services also claim that looks are not important, but that their membership is interested in "who you are."

These dating services are segregated by religion. There is Jdate, a Jewish dating service whose mission is to "strengthen the Jewish community and ensure that Jewish traditions are sustained for generations to come." Then there is Catholicmatch.com, which exhibits the pictures of men and women enrolled in that dating service and promises the protection of St. Raphael the Archangel, who is the patron saint of Catholic singles and whose Hebrew name means "God heals."

Christian dating has a large following. These dating services may be free or charge a fee for registering with them. Their advertisements rest on the premise that they open a field of contacts to the membership which is far larger than the number of people one meets in church once a week. Like all such dating services, these matchmakers tell potential members that they can be "icebreakers" for people who have a difficult time starting a conversation. Because the dating service tells seekers a great deal about other members' hobbies, education, religious outlook, and life experiences, the dating services promise that they save a lot of time, as it would otherwise take numerous dates to discover all that can be gained in a short while by the dating service. Safety is also advertised by religion oriented dating services, as it is regarded much safer than meeting strangers in public. The dating services permit users to remain anonymous as long as they choose. In addition, they provide blocking devices which prevent unwanted messages.

Religious minorities such as Buddhists, Muslims, and others also use the computer to contact others of the same faith, seeking to relate to those who share their beliefs. Then there are non-religiously affiliated dating services such as eHarmony.com.These dating services cost a subscription fee of about $40 a month for a one month membership. Those who sign up for longer periods pay less per month. Six months with eHarmony.com costs less that $30 per month, and a year costs a little less than $20 monthly.[19]

The interest in meeting those of the opposite sex is not limited to legitimate online dating. The computer has also become a vehicle for the promotion of prostitution, which is only thinly disguised as escort services. Those who advertise these services are almost exclusively women, many of whom include provocative pictures of themselves together with means of contacting them. A few men also advertise, calling themselves "sugar daddies," a reference to the money they intend to spend on the aforementioned ladies.

Pornography is added to this gender driven use of the computer. According to pornography statistics, Americans spend about $14 billion on pornography in the course of a year. This means that money spent on pornography exceeds the combined revenue of ABC, CBS, and NBC.$2.84 billion of this income is derived from the internet.[20]

In view of this vast spending on pornography, there are some who believe that those who seek out internet sexual activity are dangerous "perverts" and/or sexual predators. Considerable research has shown, however, that there is little if any connection between viewing sexually explicit material on the computer and deviant sexual behavior. Instead, such studies have demonstrated that people who have been socialized to reject sexual violence, pedophilia, or other forms of antisocial behavior, are seldom influenced to alter their conduct by reason of viewing pornography.[21]

IV.

Although violence is not a likely outcome of using computers, there is a considerable form of white collar crime which depends on the use of the computer and did not exist prior to the popularity of that device. That is called "hacking."Hacking consists of an unauthorized party utilizing a computer or an account not their own by circumventing passwords or other security measures.It is frequently used for malicious or criminal purposes and is responsible for the loss of billions of dollars from businesses so victimized. In addition, "hacking" also invades ostensibly secure government installations as well as those of private persons, and is sometimes used in new kind of crime known as cyberstalking (cyber is Greek for pilot or steerer).

Hacking is so widespread that it has been estimated that 431 million adults worldwide have been victimized by hacking or other cyber crimes. In 2011, the bill for cybercrime in the United States came to $139 billion. Worldwide, the cost of cybercrime has been estimated at $388 billion.[22]

Among many reasons for the success of "cybercrime" is the commonly held belief that computers are the product of such genius that the "average" person cannot understand enough to deal with those who know how to use computers to steal. This belief leads to the mistaken notion that the computer thief is too smart for the ordinary victim.

Hackers invade banks, every kind of website, corporations, and all kinds of government agencies. Credit card accounts have also been hacked. An example of hacking on a massive scale includes "Anonymous," who succeeded in attacking the Church of Scientology, and from there went on to attack Sony's PlayStation Network, then Fox Television, and finally the Central Intelligence Agency.These "hackers" were often students and other young people who enjoyed the anonymous hostilities they could deliver to their victims. A nineteen year old man, Jake Davis, who lives on the remote Shetland Islands, used the label "Topiary" and sent his victims such messages as "Die in a fire. You are done."A New York based hacker, Hector Monsegur, called himself "Sabu" and invaded government web sites in Tunisia.A hacker only known as William extorted pornographic pictures from his targets and then used them to split up boyfriends and girlfriends and horrify several adolescents' mothers.[23]

Although many of these "hackers" are merely interested in mischief and "fun," there are also those who seek to enter the financial records of businesses and government offices. These hackers steal financial information and medical information. They have attacked MasterCard and other corporations. The most spectacular of all these hackers is WikiLeaks, which released a large number of U.S. diplomatic cables to embarrass innumerable mindless bureaucrats. They also hacked into F.B.I. records and into private security firms and into individual bank accounts.[24]

Another computer crime which is in part victim precipitated is cyberstalking. This crime can be perpetrated on seemingly endless potential victims, as anyone using a computer may be targeted by those interested in stalking others. Opportunities for criminal victimization are produced by the lifestyle of many of the victims, as a good number of people engage in behavior which exposes them to risk. Some of the victims are victims of fraud who are generally contacted by email. These emails seek to persuade the target to send money or to reveal his bank account number or password for some invented reason. The recipient of this information then steals the victim's money or intrudes on his private life.[25]

Other victims of cyberstalking are harassed or threatened via email or chat rooms. Some cyberstalkers use other electronic devices such as cameras, listening devices, computer programs, and Global Positioning Systems attached to the victims' cars. The National Crime Victimization Study defines this kind of behavior as "the repeated pursuit of an individual using electronic or internet capable devices.[26]

College students are evidently more at risk of becoming victims of cyber stalkers than any other population. It has been estimated that 46% of college women and 32% of college men are so victimized. This is due to the common college student propensity of being connected by means of social networks such as Facebook as well as text messages, which lead to both property and sexual victimization.[27]

Females are three times more likely to be the targets of physical stalking than males and younger individuals are a good deal more likely to so victimized than older people. This is due to the greater amount of risky behavior among young men than any other group. This means that those who drink a great amount of alcohol, frequently attend parties, or spend much time in bars and nightclubs are a good deal more at risk than those who abstain from these activities. Other kinds of deviant behavior which attract stalkers are hacking, harassing others, and associating with deviant peers while online.[28]

V.

About 19 million Americans work from home by using computers. The majority of those working from home are women who have children at home. Among them are many single mothers who need to support themselves and their families. Some have very young children, while others have children already in school but who must be supervised after school and during the long summer vacations and shorter vacations and legal holidays, even when school is in session.[29]

There is some evidence that computer mediated communications weakens the emotional bonds with work colleagues. However, it also strengthens the emotional bonds with the family, particularly when fathers stay at home and do their work from there. Since the beginning of the industrial revolution in the early 19th century, "going to work" has meant leaving the home for factory or office. Prior to that time, Americans worked almost entirely from home, because 85% of Americans were farmers until the invention of the harvester, but also because non-farm work such as clothing manufacture or handicrafts of all kinds were conducted from home. The computer has partially returned employment to home work for women and for men.[30]

Men who work at home become more involved in family life than those who are absent most of the week. This results in changes in the traditional gender roles, as women can become more competitive in the labor market when the fathers of their children spend much time at home although working. Furthermore, the stereotype of the emotional woman and the stoic man is challenged when men stay home to work, because it is far more difficult to play the unemotional role when confronted all day with family interaction. For years, masculinity has been defined in terms of the man's work role. In fact, social prestige in America depends primarily on occupation. Stay-at-home men, even if fully employed, do not fit that expectation, because they no longer separate the private from the public sphere, although almost all men who work at home employ women, whether wives or "live ins" or "au pairs," to deal with children and household tasks while they work at home.[31]

Working at home involves a good number of possible distractions such as snacking, sleeping late, watching television, procrastinating, newspaper reading, drinking, or making personal phone calls. In addition, someone who works at home is subject to visits by relatives and friends and the intrusion of neighbors for numerous reasons.

In addition, there are so-called workaholics who cannot stay away from the computer even at night or on holidays or weekends. They need to work all the time and the computer at home tempts them incessantly.[32]

Work at home is not only work authorized by legitimate employers. In addition, there are numerous fraudulent "work at home" schemes which are advertised via the internet. These scams claim that the workers who reply to these promises will earn big money while staying at home stuffing envelopes, selling products, or making telephone calls. Craft assembly and other tasks are assigned to the target, who expects to be paid as an employee but receives nothing after mailing the completed task to the so-called employer. The target of the scheme pays the "company" for the materials to be assembled, such as dolls or kitchen utensils. The company, on receiving the assembled product, then claims that the assembled product "does not meet our specifications" and pays nothing. They make money from the payments received from the suckers who fell for their scheme.

There are also those who advertise, "you can turn your computer into a money making machine." This consists of paying for instructions on how the target may place ads to get work typing on the target's computer, provided he pays for a list of "employers" needing typing help. Of course, the list contains phony companies and employers who don't want the help.

Many of these schemes require that the target send money for a package that contains the product for sale. The receiver then sells the product and keeps the money. Usually the package contains worthless junk which cannot be sold to anyone.

Chain letter mailing is another scheme requiring a computer. Here the target is asked to send some money and the chain letter to the persons listed, with the promise that the chain letter will produce big money, in fact millions, to those who participate. The only people who make money are the schemers.

It is also possible to buy "medical software" which allows the victim of the fraud to solicit doctors to do their billing. This fails because doctors either have their own staff do their billing or have it done by well known entrenched companies who know a lot more about medical billing than someone who just bought a disk.

Innumerable other "work from home" frauds are constantly advertising on the internet with some success. These frauds generally target the sick, the old, the disabled, and stay-at-home mothers. Many of these people have little money but spend it on fraud in the vain hope of escaping their poverty. Almost always, the victims have little education and believe the lies sent to them.[33]

VI.

Young preschool children spend most of their time at home and therefore learn from their families what the world is all about. Consequently, children learn from their families how to deal with technology, including the computer. This means that children learn at home before they ever encounter formal schooling. Included in the outcome of dealing with computers and television at home is obesity and a failure to learn to read. Indeed, reading is first encountered in school, although prior to the introduction of technology into the American home, some children learned reading before entering school, and almost all children listened to adults reading stories from innumerable storybooks for children.

According to a Kaiser Family Foundation study, 12% of two to four year olds use a computer every day, and an additional 24 percent use a computer once a week. Older children used computers more than younger children, because the use of computers requires somewhat more motor skills and quick thinking than can be expected of younger children.

Because almost all American children live in families which own some form of technology, these children acquire operational skills such as turning on a computer or using a mouse. Children learn that taking an action leads to a response. This is generally learned as adults or older children introduce youngsters to the technological device, until the child moves on to become an independent user.[34]

The computer is of course mainly used by adults in the family, although many a schoolchild does his homework at the computer, as do college students.

The use of computers in schools has become so common that young children take for granted what was utterly unknown to their grandparents and rarely seen by their parents. In fact, the Apple computer company seeks to supplant textbooks with computers. In January 2011, that computer manufacturer introduced three free pieces of software, including what has been featured as an electronic bookstore. According to Apple, an electronic school book would cost $15 or less, which is a good deal less than printed textbooks, which cost $100 or more. Electronic textbooks accounted for only 2.8% of the $8 billion textbook sales in 2011.[35]

Although electronic books are far from commonplace in 2012, computers and their use are so ordinary in American schools that children are acquainted with computer programming, a technique at one time reserved for adults. This means that programming is now included in the elementary school curriculum. This interest in computing in grade schools is being fostered by an "Computer Science Education Week" designed to make all children computer competent. This view of grade school education has been challenged by those who believe that learning a constantly changing, ever obsolete technology is of little help in letting young children play. It is argued by a number of early childhood educators that surrounding children as young as age two with technology deprives those affected of the opportunity to be a child.[36]

One of the most unfortunate outcomes of the introduction of technology into the classroom is the belief among students and parents that the computer can be the substitute for the effort needed to master any subject matter. The evidence for this is that school performance in the United States has not improved with the increased use of computers. According to Andreas Schleicher, an official of the Organization for Economic Cooperation and Development, more and more students in other countries score higher on achievement tests than is true of American students. Testifying before the U.S. Senate Education Committee, Schleicher said that only 7 out of 10 American students get a high school diploma. These comments were confirmed by Dennis van Rockel, president of the National Education Association, and John Castellani, president of the Business Roundtable. These witnesses agreed that employers have increasing difficulty finding qualified young workers because "schools are inheriting over entertained, distracted students." [37]

Nevertheless, there are now many American schools in which children use personal computers for almost all their learning activities. Second and third graders in such schools practice arithmetic, writing, and almost everything else on personal computers. The computers used in many of these schools are supplied by Apple and other manufacturers free of charge in the hope of converting all American schools to computer use. Not only the computer manufacturers, but also politicians, have encouraged the use of computers by schoolchildren. For example, the governor of Maine asked the legislature to purchase laptop computers for every seventh grader in that state.[38]

Teachers are as much affected by the introduction of computers into the classroom as are children. Teachers change their instructional practices when using technology, as has been demonstrated by the "Classroom of Tomorrow" project. This project equipped a single classroom in each of five schools with software, printers, laser disc drives, and desktop computers, and gave each student and teacher a laptop computer for home use. These devices became catalysts for the transformation of teachers into technology users. Teachers who used computers became willing to let students teach them in areas which they did not know what students knew, organized multiple simultaneous activities during class time, assigned complex problems for students, gave students greater choice in learning tasks, and recognized student initiative in doing work outside the classroom by using computers.[39]

Another consequence of giving students laptop computers is that students will have the same opportunity to learn at home as in the classroom. This also leads to the anxiety of parents and board of education members to be on the so-called "cutting edge" of education without giving much thought to some of the negative consequences of worshipping technology.

One of these negative results of giving young students the opportunity to learn the use of computers is the harassment of teachers. This comes about because the computer age has produced impersonal and remote interpersonal relationships, and because the teaching profession is largely held in contempt by most Americans, so that children learn to view teachers with disdain.

Examples of teacher abuse by students using computers include a pedophile website in which students pretended to be the twin daughters of a vice-principal. They claimed that their father had molested them and that they had sex with their father and were now looking for a new partner.

Other students logged on to a Nazi or white supremacist website and harassed the Nazis by claiming to be a Jewish teacher in an effort to have the Nazis target the teacher. Others stole photographs from teachers' cell phones and posted them online.

"Cyberbaiting" of teachers has become commonplace in many school districts. This consists of a constant barrage of insults, humiliating messages, and embarrassing lies designed to enjoy a teacher "breaking down." Such emotional reactions are then captured on video and shown online. Students also create Facebook pages and target teachers with such comments as, "Ms. X is the worst teacher in America," etc.

All of this is a reflection of contempt for the teaching profession in the United States.

VII.

One of the most revolutionary uses of computers has been the introduction of wearable computers, used mainly in the military. These devices are worn under or on top of clothing. These devices are constantly interacting with the user and are not turned on or off but seem like an extension of the user's mind. The wearable computer has a fifty year history, beginning with the work of Edward Thorp and Claude Shannon. Shannon and Thorp became national celebrities when these Massachusetts Institute of Technology mathematicians used small computers in casinos which succeeded in using card counting techniques to beat casinos all over the world. This success was imitated by other "blackjack teams," which temporarily won a great deal of money from casinos until the players were recognized and no longer admitted to the casinos. William Poundstone wrote a book called Fortune's Formula which detailed the manner in which the M.I.T. group succeeded in defeating the casinos.[40]

Edward Thorp also applied statistical techniques to the stock market with considerable success. Subsequently, Warren Buffett, a multibillionaire, used Thorp's methods to earn vast amounts of money in the stock market.

In the 1970's, a shoe computer was devised which used radio communications to contact the bettor in a casino, and in the 1980's, Steve Mann devised a backpack mounted computer and Steve Roberts produced a bicycle with an onboard computer. Then the Student Electronic Notebook was created in 1989, followed by the wrist computer in 1994. By 2010, SONY began to sell a "smart watch," which can be converted to a wearable wristwatch computer.[41]

The U.S. military has been particularly interested in the uses of small and wearable computers. In 1994 the U.S. army launched the Land Warrior program. This program was based on interaction of the soldier with the squad leader, and depended on a computer, a radio, and a keyboard, all carried by each soldier.

The soldier wore a vest which housed the computer and other components of the electronic system. This was called the Land Warrior computer, which utilized a version of the Windows operating system with an Intel processor. The battlefield software was installed on the computer and the sensor data was downloaded on the computer.[42]

VIII.

Computers have become weapons in the effort to defend the United States against its enemies. Likewise, other countries use computers to harm the interests of this country and assault our national security, our economic condition, and our social stability. There are those who believe that the United States government should establish "red lines" in cyberspace which would be used to define the point at which the United States would defend against hostile computer driven attacks. Such thresholds, it is argued, would act as deterrents, and also insure decisive action when needed. Opponents of such "red lines" say that ambiguity as to how the United States would react to a cyber attack is more effective as a deterrent than "red lines."[43]

The number of cyber attacks on America increases constantly, so that the U.S. government has issued a publication entitled: International Strategy for Cyberspace, to the effect that the United States will respond to an attack from cyberspace as it would respond to an attack by any other means. It is now possible for a "hacker" to threaten U.S. nuclear installations, subways, pipelines, or numerous other facilities.

An example of how a cyberattack can be launched from a distance was the recent "Stuxnet" attack on the Iranian nuclear program by the United States and Israel. In 2008, at least one U.S. military computer system was penetrated by "hackers," and in 2011, Lockheed-Martin, a major military contractor, also reported having been the victim of an infiltration.[44]

To counteract these threats to national security, Congress has passed a number of laws designed to protect the nation against criminals, terrorists, hostile countries, and foreign industrial competitors. Ever since the first internet transmission took place on October 29, 1969, the United States has been forced to deal with the possible negative consequences of this invention, particularly because there are nearly two billion users of the internet. Therefore there are innumerable threats to the United States information structure.For example, in 2010, Chinese hackers attacked Google's corporate network in an effort to gain access to the company's source code. In fact, the Congressional Research Service has reported that threats to U.S. cyber and telecommunications are constantly increasing.[45]

There are a number of laws pertaining to cyber security, of which three are most comprehensive. These are the Electronic Communications Privacy Act, the Stored Communications Act, and the Foreign Intelligence Surveillance Act. Because the fourth amendment to the U.S. Constitution requires a particularized description of the place to be searched and the things to be seized, electronic surveillance by government requires legal control over such government intrusion into the private lives of citizens. The conflict between government efforts to protect the nation from enemies and the constitutional guarantees of privacy has been going on since the "tapping" of telephone and telegraph messages by federal agents began during the First World War. Later, congress passed the Radio Act of 1927, which prohibited the interception or disclosure of private radio messages.[46]

The intent of Congress was then and later to assure that wiretapping and electronic eavesdropping were subject to the fourth amendment. In 1968, Congress passed the Omnibus Crime Control and Safe Streets Act, including Title III, which prohibits wiretapping or electronic eavesdropping except for law enforcement purposes under strict supervision. Then, because of the attacks on September 11, 2001, Congress passed the Patriot Ac,t which allowed The Federal Intelligence Surveillance Court to use electronic surveillance for intelligence and enforcement objectives. These laws are not clear as to whether everything the security community does on computer networks is subject to electronic surveillance.[47]

IX.

Computers have altered American life drastically. Online shopping, entertainment, banking, travel, education, and meeting people are only a click away. Everyone who has access to a computer also has access to more information than was ever possible before. It appears that everything is to be found on computers. This therefore means that knowledge is available to everyone, and not only to learned or educated people. Limited only by availability of communication networks and electricity, the world is open to all, wherever they may be. While erstwhile the reader of books and magazines or newspapers was the passive recipient of information limited to the content of the literature he was reading, the computer is interactive and allows the reader to respond. Distance and time are no longer barriers, as people can reach anyone anywhere.

Mobile phones have increased communication in a dramatic manner, as people can talk to anyone at any time, located anywhere on the globe, as handheld phones have become commonplace. These phones have also given rise to two new traffic hazards, phoning and texting while driving. In addition, automatic teller machines have made banking nearly obsolete, as money can be obtained at any time, day or night, even when banks are closed.

On the negative side, we no longer have the choice of talking to a live person on a telephone at will. Instead we are confronted with innumerable number options used by businesses, even at doctors' offices, and mindless bureaucrats of all sorts, including every kind of government official. The most egregious outcome of the electronic revolution is government surveillance of citizens and their daily activities. Police have installed digital cameras all over cities and are using computers to gradually reduce the United States to a police state in which "the land of the free" is becoming not much more than a phrase in a song.

Microwave ovens, answering machines, airbags, and computers are part of daily life, even as new electronic devices are sold incessantly. We even have digital cameras. Credit card and other financial activities leave a computer trail used by police and criminals to threaten the financial integrity of just about anyone.[48]

The current cyber revolution is by no means a new experience, in that European history clearly shows how the invention of the printing press had similar outcomes for those living in the 15th and 16th centuries. Johann Gutenberg was the first European to create a printing press in 1440, although a Chinese printer had invented a moveable clay type in 1041. Because Chinese symbols are so numerous, the Chinese method was very cumbersome and produced few books. Nevertheless, printing with movable type is a Chinese invention.[49]

The cyber revolution has allowed the retired population to overcome the loneliness which American culture dictates must be the role of those deemed old. The same population is also using the computer to access medical information, although this risks self doctoring, which may have negative consequences.

All of this refers to the insight of Marshall McLuhan who, in the 1960's, predicted the arrival of "the global village," by which he meant that an electronic nervous system known as the media would integrate the entire planet, so that events in any part of the world would impact all other parts of the earth.

This prediction has come true, as mankind now lives once more in a small village, albeit the entire globe.[50]

Because all human inventions and discoveries have been subject to all kinds of fraud, this is also true of computers and their derivatives. "Cyberstalking," electronic plagiarism, cyber-pedophilia, cyber-pornography, outsourcing, and other negative outcomes are associated with the global village constructed by the internet. That internet has affected the manner in which we organize our time. A click can get us what we want in a millisecond, saving all the time it would have taken to run to a store, find what we want, and wait in a checkout line to get back to our car and drive home. We don't even need to get dressed. We can do it all on the computer while loafing in our underwear.[51]

Currently, researchers are studying how to combine the human brain with technical instruments such as the computer. A company called Applied Digital Solutions has developed an implantable "verichip" which can be inserted into a heart valve so as to discover that the patient has become unconscious; or the chip can be inserted into eyeglasses and broadcast the location of the person wearing them. Children and even pets can be supervised at all times by inserting a chip into them. So much can be done to influence human behavior with electronic devices that the difference between augmented reality in the analog world becomes welded into the digital world. Finally, the electronic world and the computer will tell government where all citizens are at any time and what they have read, said to others, bought in a store, and perhaps even what they are thinking. We are on our way to the well ordered police state and the computer is its forerunner.[52]

Summary

Computers are the product of many years of effort by numerous inventors, including Pascal, who invented the first working calculating machine. In the 21st century, computers are widely used in the United States and around the world. Texting, email and computer games are now commonplace, as is hacking and cyberstalking.

The computer has reintroduced work from home in America, even as children at the youngest age learn to use computers at home and in school. This achievement has some negative consequences, such as computer driven harassment of teachers and students.

Computers have made it possible to beat casino games and produce wearable computers for military use. National security is threatened by hackers ,so that numerous laws have been passed by Congress to defend against such threats from America's enemies, even while protecting the fourth amendment.

In sum, the computer has made the world a "global village," as further advances in electronic research promise to lead to the total control of the individual by government through the insertion of "chips" into the citizens, even of erstwhile free countries.

Notes

1. Carl B. Boyer, *A History of Mathematics,* (New York: Wiley, 1991), pp. 252-253.

2. Henry C King, *A History of the Telescope,* (New York: Dover Books, 2003), pp. 114-118.

3. Betty Mayfield, "Gerbert d'Aurillac and the March of Spain: A Convergence of Cultures," *The Mathematical Sciences Digest, (no date),* p. 6.

4. Samuel Chapman, "Blaise Pascal Tercentenary of the Calculating Machine," *Nature, (October 31, 1942),* vol. 150, pp. 508-509.

5. Samuel Chapman, "Blaise Pascal etc.," pp. 508-509.

6. Michael Dobbs, "IBM Technology Aided the Holocaust," *The Washington Post,* (February 11, 2001), p. A22.

7. No author, "Alan Turing," *Time Magazine,* vol. 154, no. 27 (December 31, 1999).

8. Georges Ifrah, *The Universal History of Computing,* (New York: John Wiley and Sons, 2001), pp. 95, 207-210, 217-220, 282-283.

9. Robert Cailliau, "A Short History of the Web," *Net Valley,* (November 2, 1995), p. 1.

10. *Occupational Outlook Handbook,* (Washington, DC: U.S. Bureau of Labor Statistics, 2012-2013).

11. Benjamin Pierce, "*Types and Programming Languages,* (Cambridge, MA:. The MIT Press, 2004), p. 339.

12. No author, "Roxxxy Sex Robot World's First Robot Girlfriend Can Do More Than Chat," *Huffington Post,* March 18, 2006), Technology, p. 1.

13. Mark Hachman, "Facebook Now Totals 901 Million Users," *PCMagazine,* (August 23, 2012), p. 1.

14. Cedric Price, "How Skype Works," *Ezine Articles,* http://exinearticles.com.

15. Mark D. Griffiths, "Videogame Addiction," *International Journal of Mental Health and Addiction,* vol. 6, (2008), pp. 182-185.

16. Ng, B. D. and Peter Wiemer-Hastings, "Addiction to the Internet and Online Gaming," *CyberPsychology and Behavior,* vol. 8, no. 2, (2005), pp. 110-113.

17. Samuel Fisher, "Identifying Video Game Addiction in Children and Adolescents," *Addictive Behavior,* vol. 19 (1994), pp. 545-553.

18. Christopher J. Ferguson et. al., "Violent Video Games and Aggression," *Criminal Justice and Behavior,* vol. 35, no.3, (March 2008), pp. 311-332.

19. No author, "Online Dating Guide and Blog: What is the Cost of eHarmony Subscriptions?"
http://www/littleredrails.com/blog/what-is-the-price-of-eharmony-subscriptions/.

20. No author, "Pornography Statistics," *Family Safe, Media,* (2012), p. 6.

21. Neil M. Malamuth, et. al., "Pornography, Individual Differences in Risk, etc." *Sex Roles,* vol. 66, (2112), p. 427.

22. Jen Weigel, "Cybercrime: A Billion Dollar Industry," *Chicago Tribune,* (September 26, 2011), p. 3.

23. Janet Maslin, "The Secret Lives of Dangerous Hackers," *The New York Times,* (June 1, 2012), Sec. C., p. 27.

24. Bill Saporito, "Hack Attack," *Time,* vol. 178, no. 1, (July 4, 2011).

25. Majid Yar, "The Novelty of Cybercrime," *European Journal of Criminology,* vol. 2, (2005), pp. 407-427.

26. Bradford W. Reyns, "A Situational Crime Prevention Approach to Cyberstalking Victimization," *Crime Prevention and Community Safety*, vol.12, (2010), pp. 99, 118.

27. Bonnie S. Fisher, et. al. *"Unsafe in the Ivory Tower,* (Thousand Oaks, CA: Sage, 2010), pp. 160-163.

28. Elizabeth E. Mustaine and Richard Tewksbury, "Predicting Risks of Larceny Theft Victimization," *Criminology* , vol. 36, (1998), p. 829.

29. Adrienne Samuels Gibbs, "Make Working from Home Work for You," *Ebony*, vol. LXV., no. 12 (October 2010), p. 1.

30. Stephen Fineman, "Getting the Measure of Emotion - and the Cautionary Tale of Emotional Intelligence," *Human Relations*, vol. 57, no.6, (2004), p. 719.

31. Melanie Wilson and Anita Greenhill, "Gender and Teleworking Identities in the Risk Society," *Work and Employment*, vol.19, no. 3, (2004), p. 207.

32. Kathy Koch, "Flexible Work Arrangements: Do They Really Improve Productivity?" *CQ Researcher*, vol.8, no. 30, (1998), p. 697.

33. Audri and Jim Lanford, "Home Based Business Scams," *Internet Scams*, no. 61, (June 2012), p. 1.

34. Lydia Plowman, Joanna McPake and Chistine Stephen, "Just Picking it Up: Young Children Learning with Technology at Home," *Cambridge Journal of Education*, vol. 38, no. 31, (2008), pp. 303-319.

35. Brian X. Chen and Nick Wingfield, "Apple Introduces Tools to Supplant Print, " *The New York Times*, (January 20, 2012), BITS 3.

36. Greg Simon, "Sunday Dialog: Using Technology to Teach," *The New York Times*, (October 20, 2011), Letters, p. 2.

37. Sam Dillon, "Many Nations Passing U.S. in Education, Expert Says," *The New York Times*, (March 10, 2010), p. A21.

38. Mary Ann Zehr, "Laptops for All Doesn't Mean They are Always Used," *Education Week*, vol. 19, no. 30, (2000), pp. 14-15.

39. Henry J. Becker and Jay Ravitz, "The Influence of Computer and Internet Use on Teachers' Pedagogical Practices and Perceptions," *Journal of Research and Computing in Education,*vol. 31, no.4, (1999), p. 350.

40. William Poundstone, *Fortune's Formula*, (New York: Hill and Wang, 2005), pp. 41-43.

41. Kevin Warwick, *I, Cyborg*, (Urbana: The University of Illinois Press, 2004).

42. Matthew Cox, "Troops in Iraq give Thumbs Up to Land Warrior," *Army Times*, (June 23, 2007), p. 1.

43. John A. Mowchan, "Don't Draw the Red Line," *U.S. Naval Institute Proceedings*, vol. 137, no. 10, (October 2111), p. 1.

44. Siobhan Gorman and Julian E. Barnes, "Cyber Combat: Act of War," *The Wall Street Journal*, (May 30, 2011), Technology, p. 1.

45. Bill Gertz, "Inside the Ring: PLA Hack on Google?" *Washington Times*, (July 8, 2010), p. A7.

46. *Radio Act of 1927*, Public Law no. 632, Chapter 169, 44 Stat. 1172.

47. U.S. Patriot Act, (2001) Public Law no. 107-56115.Stat.

48. Carmen Luke, "Buzzing Down the On-ramp of the Super-highway," *Social Activities*, vol. 20, no. 1, (January 2001), pp. 1-8.

49. http://www.ideafinder.com/history/inventions/printpress.htm.

50. Eric McLuhan, "Marshall McLuhan Forsees the Global Village," *Marshall McLuhan Studies*, vol. 1, no.4, (May 16, 2000), p. 13.

51. Carmen Luke, "What next? Toddlery Netizens, Playstation Thumb, Techno-literacies," *Contemporary Issues in Early Childhood*, vol.1, no. 1, (1999), pp. 95-100.
52. Miriam Meckel, "Mensch wird Maschine," *Die Zeit*, (June 12, 2012), p. 13.

Epilogue

Technology and Social Change

I.

Since the introduction of the steam engine into American life at the end of the eighteenth century, technological developments have altered the nature of the U.S.A. radically. In 1790, when the first census was taken, the population of the United States was reported as 3,893,635. This is undoubtedly an undercount of the people living in the thirteen original states, because technology did not allow for a more accurate accounting, because some people refused to be counted, and because the distances between communities were too great to allow census takers to visit everyone. Although no statistics concerning occupation are available for the 1790 census, historians have estimated that 90% of Americans lived on farms at that time.[1]

The preponderance of agricultural employment in the United States continued until 1910. Throughout those years, the United States exported farm products and imported technological devices, mostly from England. As the frontier expanded after the Civil War (1861-1865), the principal occupation of farmers was the clearing of land, as cotton became the main income producer for the slave economy in the South.[2]

By contrast, the American population had become so urbanized after the two world wars that by the 1980's the farm population in this country had shrunk to less than two percent, or 4,986,000 out of a population in 1986 of just over 241 million. Then the 2007 Census of Agriculture revealed that there were then only 2.2 million farmers in the United States when the population of the United States had increased to over 301 million. Of the 2.2 million farmers, 1.2 million, or 65%, were earning additional income from non-farm occupations. This then is the most dramatic consequence of technological inventions since the middle of the nineteenth century.[3]

As the population of the United States grew, slower growing sectors of the economy lost some of the resources available, as more rapidly growing sectors expanded with the aid of new technology, leading to new products and new services. This was already visible in the 19th century, but became impressive after the First World War, i.e. during the 1920's. Then more and more Americans owned cars and bought new household appliances and used electricity, which in turn drove growth. For example, electric lights, radios, electric iceboxes, electric fans, and electric irons all came into widespread use. It was in 1913 that the assembly line was invented, which increased productivity in the 1920's, even as the workweek began to decline in a few industries.[4]

One means of recognizing the impact of new technology on American life is to compare the gross domestic product as it grew over the years. In 1820 all the goods and services consumed by an individual American amounted to $1,257. By 1870 this had grown to $2,245, and by 1913 it reached $5,301. In 1950 the gross domestic product consumed by an average American had attained $9,561 and at the end of the 20th century it reached $27,235.[5]

As the gross domestic product increased, life expectancy in the United States also increased. For both sexes, the average life expectancy at birth in 1850 was 38.3 years. By 1860, that had increased to 41.8 years, and by 1900 it reached 47.8 years. Forty years later, in 1940, life expectancy in the United States stood at 62.9, years and in 1998 it had reached 76.7 years. The Bureau of the Census estimates that life expectancy for both sexes will reach 78.9 years in 2015. No doubt the agricultural revolution, as well as advances in public health and medicine, have made this increase in longevity possible.[6]

As Americans ate more meat and other proteins, the average height of men and women increased. At the outset of the 21st century the average American eats about 200 pounds of meat, 33 pounds of cheese, and 60 pounds of other fats and oils each year. A century earlier, Americans ate only 125 pounds of meat a year. Prior to that, meat was served twice a week in American households. As a result of this increase in consuming proteins, Americans have grown taller and heavier in one century.[7]

In 1800 the average height of American men was 68.1 inches. In 1930 this had increased to 69.2 inches, and by 1979 the average American man had reached 69.8 inches, or almost five feet ten inches. For women, height measurements were not available before 1930 because women did not serve in the armed services of the country, where male height had been measured. In 1930, the average American woman was 64 inches tall. In 1970 this had increased to 64.4 inches and remained there since. The weight of American men increased from 166 pounds in 1960 to 191 pounds in 2002. For women, the average weight in 1960 was 140 pounds, but increased to 164 pounds by 2012.[8]

In view of this increase in the height and weight of Americans it is noteworthy that menarche has also been affected by the increase of the food supply and the greater consumption of proteins. While the onset of menstruation for American women was approximately 17 years in 1890, a gradual decline has led to menarche at about age 13 in 2010. Meanwhile, the median age at marriage for American women has increased from 20 years in 1950 to 26 years in 2010. In earlier years, the median age at marriage was yet lower. As a consequence, the pregnancy of unmarried and never married mothers has increased dramatically in the United States. During the first decade of the 21st century, about four out of ten American children were born to unmarried mothers. It is evident therefore that the increase in unmarried motherhood is at least in part due to the increase of time between menses and marriage in this century.[9]

II.

In the second century when Trajan and his successor Hadrian were emperors of Rome, the empire included all the lands bordering on the Mediterranean Sea and extended through England all the way to the border with Scotland. The empire encompassed one million nine hundred thousand square miles, or about two thirds of the forty-eight United States outside of Alaska and Hawaii, and had a population of about 88 million people. The city of Rome was the largest in the world at that time, with a population of one million.[10]

The emperors held this vast empire together by force of arms, as they were unable to speak to all these people unless they traveled incessantly to various parts of the empire. Even then, they could only address those few within distance of their voice.

Now, in the 21st century, the president of the United States, or anyone who has access to the American system of communication, can address millions of people living across the country from Hawaii to Puerto Rico and across 3,800,000 square miles of the land. Evidently, communication has changed the United States from its eighteenth century roots to a unified power with one government, one language, one culture, and one history. Indeed, there are numerous ethnic groups living in the United States in the 21st century who may differ in their language of origin and in the origin of their ancestry. Nevertheless, mass communication has made it possible to talk to Americans in the same idiom from California to Vermont, as Americans listen to the same music, are harangued by the same politicians, and laugh at the same jokes at the same time all over the country.

When the third president of the United States, Thomas Jefferson, induced Meriwether Lewis and William Clark to explore the route from St. Louis to the Pacific, it took the fifty-nine people associated with the expedition from May 14, 1804 to November 1805 to reach the Pacific Ocean. Only thirty-three members of the expedition went all the way. The expedition returned to St. Louis on September 23, 1806. They had traveled 3,700 miles on a trail to the ocean and seven thousand miles after returning. Today the seventeen hundred miles between St. Louis and Portland, Oregon are traveled in hours. That then is one of the most significant consequences of technology.[11]

Once the Pacific Ocean had become a reality for Americans by reason of that expedition, thousands and finally millions traveled west from St. Louis, Missouri to the west coast with their whole families in covered wagons. The first train of these wheeled vehicles left that city on April 10, 1830. These pioneers traveled from 15 to 27 miles a day over turbulent rivers and steep riverbanks and even the Rocky Mountains.[12]

This meant that the frontier became the defining event in American history and culture, as shown by Frederick Jackson Turner in his epochal essay of 1893, which showed that as of the census of 1890 there was no frontier in America as the land all the way to the Pacific had been settled. According to Turner, the emigration west created a new people and new institutions. Turner called Americans a "mixed race," a term now rejected in favor of "ethnic groups." This meant to Turner that newcomers from anywhere were integrated into the new population, leading to true democracy, as anyone without reference to origin or social standing could participate. In short, the western frontier secured democracy in this country because "free land" allowed economic opportunity to anyone willing to move west. The frontier also led to individualism in the U.S.A., and because of the distances from Washington, D.C. to the new settlements, to a distrust of government and to self reliance as an American virtue.[13]

III.

Immigration to the United States began as a steady stream shortly after the English colonists had declared their independence from the mother country, England, in 1776. Yet, the number of immigrants who came here after the revolution was relatively small when compared to the size of immigration after 1867. The reason for this sudden increase at that time was the employment of the steamship, which was gradually replacing sailing ships in making that long transatlantic voyage. For example, between 1780 and 1819, only 9,900 immigrants came to this country, and between 1820 and 1830, 14,538 immigrants arrived. This increased to 71,916 between 1832 and 1846, but was vastly exceeded between 1847 and 1854, when334,506 immigrants came, almost entirely from western Europe.[14]

The reason for this sudden increase was the introduction of the steamship into the North Atlantic route from Europe to North America. Before the Civil War (1861-1865) most immigrants arrived by sail, and after that era most immigrants arrived by steam. It was the invention of the steamship which made the increase of immigration possible. In 1852, fewer than 20 steamships arrived with about forty immigrants, but in 1867, 47,000 immigrants came to New York on sailing ships.[15]

As the steamers became larger and faster, a flood of immigrants from Europe arrived here. It has been estimated that between 1860 and 1924, twenty-five million people crossed the ocean on steamships. The steamship allowed migrants to either seek new homes or temporary employment followed by a return to Europe, because the voyage was far faster than was ever possible by sailing ship. Thus, Columbus traveled three months from Spain to Hispaniola, although later sailing ships reduced that time to five or six weeks. The steamship took two weeks or less, which also reduced mortality on board. This meant that those who had come by sail could hardly expect to return after working in the U.S.A. temporarily. The steamship made that possible, although the vast majority of immigrants traveled in a part of the steamship called "steerage" into which multiple rows of beds were crammed. These ships had inadequate ventilation, too little food and hardly any sanitation. The immigrants therefore arrived hungry, dirty, and exhausted.[16]

Nevertheless, they came by the millions, not only because they were expelled from their homelands by poverty, war, and tyranny, but also because technology had made such a mass migration possible.

One more consequence of the rise of technology in the nineteenth century was the growth of American cities. Immigrants, as well as native born Americans, entered American cities in vast numbers after 1870 to take advantage of the technological inventions which seemed to increase economic security over the uncertainties of farming, and which sought cheap labor as productivity in urban factories increased.

In colonial days, city life was nearly unknown in this country, as only 3.35% of Americans lived in small towns such as New York and Philadelphia. By 1840, the proportion of the non-farm population had grown to 8.52%. Then, in 1870, over twenty percent of Americans lived in urban areas, and by 1890, more than 62 million Americans lived in 448 cities, constituting nearly thirty percent of the population.[17]

The conditions under which the immigrants worked and lived resembled the slavery which had been abolished in this country with the 13th Amendment to the Constitution in 1865, but continued in the guise of sharecropping for many years thereafter. Both in the South and in the North, semi-slavery remained. Yet, it is reasonable to hold that plantation slavery as practiced in the southern states had to disappear even without the Civil War, because technology would have put an end to it. The evidence is that slavery never took hold in the northern states because the climate did not lend itself to a plantation economy, but also because technology made slavery unprofitable. Evidently, new technology and new factories made investments much more attractive than southern agriculture. It is fairly certain that slavery would have survived somewhat longer had the Civil War not put an end to it. Yet, it is also evident that the subsequent sharecropper system continued slavery in a milder form until the south caught up with the north in technological productivity.[18]

Sharecropping came about because the freed slaves had no money. The majority could not afford to move north because they had families. Therefore they accepted a system whereby the farmer worked the land of the owner, who then collected one half to two thirds of the profit. This kept the farmer in perpetual poverty and resembled the slavery just overcome.

Meanwhile, the huge immigration of Europeans to the northern states and their factories led to the semi-enslavement of these poor foreigners, so that technology both reduced slavery in the south and increased it in the north, albeit by another name. This was depicted by Jacob Riis (1849-1914), a Danish immigrant who became a reporter and photographer in the early twentieth century. His book How the Other HalfLivesaroused the public to the heinous conditions of semi-slavery in which the poor existed in hideous slums and crippling conditions in American cities.[19]

Anyone who has read Upton Sinclair's (1878-1968) book The Jungle, first published in 1905, is acquainted with the horrendous conditions immigrant workers faced in the Chicago slaughterhouses at the beginning of the 20th century. Similar horrors were imposed on the newcomers in other industries as well, so that wage slavery existed in the United States from coast to coast. Technological slavery gradually gave way to the electronic age and to legislative efforts seeking to improve the work conditions of American employees. Yet it is evident that technology can be both the friend and the enemy of labor, a verity which translates into the electronic age as well.[20]

IV.

The inventions and discoveries of the nineteenth century led directly to the promotion of American capitalism and to agitation against capitalism. The word capital is related to the word chattel, or cattle, meaning head. The Bible records that the patriarch Jacob was a wealthy man because he owned much cattle (Genesis 30:43), a condition which even today insures great wealth among Texas cattle ranchers. The Romans collected a head tax from each citizen, which was called capita. Then, in the 19th century, the founder of communism, Karl Marx (1818-1863), used the German word "Das Kapital" in his epochal work, published in 1867.

Nineteenth century capitalism has been greatly altered over the years. Yet its essentials still exist in the 21st century. Accordingly, the capitalist is the owner of the means of production, consisting of land or factories and the machinery installed there. Karl Marx wrote that the owner buys the labor of those who have only their labor to sell. This makes the owner an employer who buys labor in exchange for wages. The workers who agree to work for the employer have no access to wealth and therefore no means of resisting the unbridled greed of the employers, who paid as little as possible and gained as much as possible from the misery of the helpless laborers. According to Marx and other observers of the 19th century capitalist system, the employer made a profit by exploiting the labor he has bought in that he pays less than the worth of the finished products he sells. Marx contended that the capitalists, whom he called "the bourgeoisie," a French word meaning citizen, was exploiting the "proletariat," a Latin word meaning "many children," from "proles," i.e. child. Since those who have many children are likely to be poor for that reason, Marx employed this phrase. "Das Kapital" holds that the value of a finished product is the labor which produced it, and that therefore the workers should be paid the entire value, with nothing going to the capitalists. This was to be achieved by creating a dictatorship of the proletariat.[21]

Some of the most angry assaults on technological developments and capitalism came from those Americans who followed Karl Marx and his writings. This means that Marx had a great deal of influence on the radical anti-capitalism movement in the United States at the end of the 19th and the beginning of the 20th century. The unquestioned leader of the anti-capitalist movement was Emma Goldman (1869-1940), who preached her message of defiance of the great employers in this country. She did so without any technical support and with a great deal of physical and emotional investment. Goldman had emigrated from Russia. Together with others of similar convictions, Goldman sought to overthrow all government, and therefore preached anarchy to whatever audiences she could find among the laborers and immigrants in the New York area. On August 23, 1893, she spoke to 3,000 people in Union Square in New York, telling the unemployed to take action. Imprisoned for incitement to riot, Goldman traveled to Europe after her release, where she met other anarchists who believed that the advances of technology and the connivance of governments led to the destruction of the laboring classes.[22]

Neither the anarchists nor the unbridled capitalists won the arguments of the nineteenth century. Instead, as more and more technological inventions flooded America, organized labor prevailed to limit the power of employers. This came about as government during the Franklin D. Roosevelt administration introduced numerous safeguards securing a better life for the workers in America while at the same time supporting capitalism, although in altered form.

V.

Technology liberated women from the age old dependency that had for centuries reduced them to second class humans. Ever the victims of the slogan "barefoot and pregnant," it wasn't until technological developments made it possible to avoid incessant pregnancy that women finally achieved equality with men and succeeded in overtaking some American men in education and income in the 21st century. This was made possible when "the pill" finally became a secure method of birth control.

Before 1840, marital fertility in the United States for the white population was 6.6 children per woman. By 1900, when the population of the United States had reached 75 million, the fertility rate had dropped to 3.6 children per woman. The growth of the American population was largely due to highly fertile migrants, as one and one half million immigrants came to the U.S. every year between 1902 and 1910.[23]

In 1900, only 20% of women were in the American labor force. At the outset of the 21st century, 60% of women had entered the labor force. Seventy percent of these women work in white collar jobs and 57% are entering the professions. This has become possible because technology has largely eliminated muscle work and provided men and women with work which requires some education but not much physical strength.

Until the 1940's, the majority of employed women were engaged in factory labor or office occupations. By the 21st century, only 26% of women were employed in factory of similar work. Technology has also relieved women of many of the traditional tasks that occupied all of their time all their lives. When most Americans lived on farms, women had to deal with an average of six children. They were constantly employed in food preparation, which occupied most of the day. Then there was no refrigeration, no packaged food, and no frozen dinners. Furthermore, so few women could work outside the home that families had only one income, while technology has permitted the reduction of the family to 1.9 children and two incomes.[24]

Labor saving devices also impinged on the servant occupation. Instead, services were obtained outside of the home, where dry cleaning, shoe repair, ready to wear clothes, baking, and numerous other services could now be bought in stores.[25]

Technology has also increased the proportion and numbers of American women achieving a higher education. While 63% of college students in 2012 are women, this was far from true forty years earlier, when only 43% of female high school graduates attended college. In 1960, only 5.8% of women of college age graduated from college. This has risen dramatically by 2012, as more women than men earned bachelor's degrees in science and engineering than was true of men in that year.

VI.

The influence of technology on religion has been debated for years without resolution.

If church attendance is any indication of an interest in religion in America, then the frequency of such attendance is an indication of the influence of technology on American secularization. According to a 2010 Gallup poll, approximately 43% of Americans attend religious services once a week. This contrasts with attendance at weekly religious services in the nineteenth century and the first part of the twentieth century, when such attendance was far greater. Nevertheless, it is a matter of historical record that many of the early Americans, other than the Puritans, did not participate in religious activities, although technology had not been developed yet.

A Gallup Poll of 2010 also shows that twenty percent of Americans never attend any religious functions and 25% attend seldom. Therefore it is reasonable to assume that only a little more than half of all Americans have an interest in established religions. Since private religious interests cannot be measured, the results of the polls are assumed to be reliable.[26]

When Andrew Dickson White (1832-1918) published his monumental History of the Conflict between Religion and Science in 1874, many educated Americans believed that technology would gradually eliminate religion from America altogether and that science would triumph in the end. That was also the view of John Dewey (1859 - 1952).Dewey was the father of American humanism and a convinced atheist whose influence on American students was indeed profound. Yet, White did not prevail in the late 20th century despite the Galileo affair (1614) and the Scopes trial (1925), because these and similar incidents could not defeat all religion in the long run. This means that at one time enthusiasts for science believed that science would eventually explain all things to all men. This has not happened, as science deals in probability but has no final answers. Religion has final answers but ignores the lessons of probability. Hence there is a standoff but no final resolution to the presumed conflict.[27]

When the Gallup poll asked Americans in the 1970's whether religion can answer their problems or is out of date, seven percent of those polled held that religion was "old fashioned." In 2010, over 28 percent believed that religion is out of date. A similar poll asked Americans whether they have confidence in organized religion. In 1973, sixty-five percent of respondents answered in the affirmative, but only 44% thought so in 2010. Americans also believe that religion is losing its influence. While 69 percent of Gallup poll respondents thought religion was influential in 1957, this dropped to 27% in 2008. Finally, it is significant that in 1948 only four percent of Americans had no religious identity, a number which rose four times, to 16%, in 2008.[28]

It is popular and commonly believed that religion has gradually given way to science and that we can expect the end of religion as science and technology solve all of mankind's problems. This belief has been refuted by the evidence that religion is not becoming less important in our technological society. While ideology and prejudice among some academics has for years supported the contention that religion is in decline, empirical evidence suggests otherwise.[29]

Although Emile Durkheim (1858-1917) may well be considered the father of the sociology of religion, his view that religion is merely a carryover from humanity's primitive past is outdated, outmoded, and proven wrong. Influenced by that estimate, a number of "intellectuals" predicted the end of religion altogether, preceded by a severe decline in a short time.[30]

The evidence shows that religions which focus on the "next world" seem to be more successful in attracting members than is true of religions focusing on this world. For that reason, the so-called 'liberal' denominations are losing members to the conservative denominations, which have grown considerably at the expense of "mainline" denominations. Therefore, religion is not experiencing a decline but rather a shift from some denominations to other denominations, based on how these groups can meet the needs of the population.

VII.

American technology did not stay exclusively in the United States. Instead it was exported to the entire globe and has influenced the world's population profoundly since the nineteenth century. Many of the exports of American technology were and are of a military kind, so that it is visible that numerous countries around the world are heavily dependent for their defenses on American technology. This came about as a result of World War II. Prior to that war, America did not have a large defense industry. After that war, the contrary was the case, as American war materials dominated the world. The production of American military technology also won the so-called "cold war" against the erstwhile Soviet Union. Thereafter, military spending declined until the attacks of September 11, 2001, which once more energized the military technology industry. There can be little doubt that the speed of technical progress in the United States has been strongly influenced by the worldwide arms race which made military production so profitable.[31]

American technology was greatly expanded by reason of military expenditures. Military needs created an extensive network of government run laboratories and the involvement of universities and their faculties such as the Massachussetts Institute of Technology, Caltech, Bell Laboratories, and others. This expansion of military technology was accelerated with the development of the Intercontinental Ballistic Missiles project, which started in 1954, and subsequently by the Apollo project, which rocketed into space in the 1960's.[32]

After the United States defeated the Soviet Union in the space race, the National Aeronautics and Space Administration was continued by promoting the space shuttle, resulting in developments of heat resisting materials, high speed computers, microelectronics, and photography. All of this and more influenced civilian technical progress and led to the diffusion of military technology to civilian use. Aerospace, telecommunications equipment and pharmaceuticals all benefited from military research. The internet was created by the military but was made widely available to civilians. Civilian startup firms developed as a byproduct of military technology. Thus IBM benefited greatly from government sponsored research, as did AT & T in digital data communications, and other companies in science based industries.[33]

Because the United States is home to 139 of the world's largest companies, 20 million small businesses, and most of the world's billionaires, American technology has dominated the world economy for some time. It was American technology which gave the world the internet, the cell phone, and air conditioning. In addition, automobiles, electronics, and aircraft used around the world come mostly from America. Financial services are also exported to the whole world by means of the New York Stock Exchange, the largest stock exchange in the world. That stock exchange is largely dependent on technology. The United States is also the largest exporter of agricultural products such as tobacco, wheat, and corn, all produced by agricultural technology.[34]

As globalization has expanded and American technology has been diffused all over the world, American ideas, beliefs, and political conditions have come to influence many of those who use such technology. The television, radio, internet, and movies all come form America and project American culture to every part of the globe.

This in turn has led to the disintegration of many dictatorships all over the world, particularly in the Muslim world, where several potentates have been overthrown by their own people. The masses of people who revolted against their oppressors are influenced by the information revolution created by the computer and other devices. As more and more people use the internet, resentment against political hierarchies is fueled by connection to democratic values coming directly from America and other free societies. Globalization of technology is therefore the catalyst for globalization of democracy.[35]

It is telecommunications which has spread the wish for freedom around the world and altered the most traditional hierarchical political systems. The best example of this trend was the defeat of tyrannical communism in the erstwhile Soviet Union and the revolt against a number of dictators in the Middle East. In all these cases, capitalism and democracy triumphed over repression, as the "authorities" were no longer able to limit access to information. Central control of all citizens has become almost impossible in the information age, so that even communist China has turned to capitalism to survive. Networks are increasingly powerful in the technological society, similar to the already existing telephone network, which encompasses the entire globe. [36]

Physical labor is almost gone in high tech societies. It is replaced by intellectual work due to the information revolution. Consequently, education and continuing training will be essential as rapid change will make it impossible to rest on yesterdays' achievements.

As globalization grows, borders between countries will become less and less meaningful. Movies, news, television, and inexpensive telephone calls as well as email have begun to homogenize younger people whose interest in music, sports, and entertainment begin to resemble one another all over the world. The same sports icons, the same actors, the same singers and performers, are now available to everyone everywhere across the continents and around the world. Brand names, mostly made in America, like Coca Cola, McDonalds, Disney, and Pepsi, have become ubiquitous as the world shrinks and the "global village" has finally arrived.[37]

Notes

1. Virginia D. Anderson, "Thomas Minor's World: Agrarian Life in 17th Century New England," *Agricultural History,* vol. 82, (Fall 2008), p. 496.

2. Chad Montrie, "Men Alone Cannot Settle a Country: Domesticating Nature in the Kansas –Nebraska Grassland," *Great Plains Quarterly,* vol. 25, no.4, (Fall 2005), p. 245.

3. U.S. Department of Agriculture, "2007 Census of Agriculture," http://www.agcensus.usda.gov.

4. Ross M. Robertson, *History of the American Economy,* (New York: Harcourt, Brace and Co., 1955), pp. 216-240.

5. Angus Maddison, *The World Economy: A Millenial Perspective,* (Paris: OECD 2001), p. A-1c, A1-d.

6. U.S. Department of Commerce, Bureau of the Census, "Expectation of Life at Birth," *Statistical Abstracts,* (2012):104.

7. Neal Barnard,"Do We Eat Too Much Meat?" *The Huffington Post,* (January 12, 2011), p. 1.

8. Robert Longley, "Americans Are Getting Taller, Bigger, Fatter, Says CDC," *U.S. Government Info,* (October 28, 2004), p. 1.

9. Robert Stein and Donna St. George, "Unwed Motherhood Increases Sharply in U.S." *The Washington Post,* (May 14, 2009), p. 1.

10. Will Durant, *Caesar and Christ,* (New York: Simon and Schuster, 1944), pp. 407-419.

11. Donald Jackson, "The Public Image of Lewis and Clark," *The Pacific Northwest Quarterly,* vol. 57, no. 1, (January 1966), pp. 1-7.

12. Albert Hawkins, "Centennial of the Covered Wagon," *Oregon Historical Quarterly,* vol. 31, no.2, (June 1930), pp. 115-124.

13. Martin Ridge, "The Life of an Idea: Significance of Frederick Jackson Turner's Frontier Thesis," *Montana: The Magazine of Western History,* vol. 41, no.4, (Winter 1991), pp. 2-13.

14. U.S. Bureau of the Census, Population Division, "Immigration Volume and Rates, (March 9, 1999), Table 4.

15. C. Knick Harley, "The Shift from Sailing Ships to Steamships, 1850-1890: A Study in Technological Change and Its Diffusion," In: *Essays on a Mature Economy,* Donald A. McCloskey, Editor (Princeton, NJ: Princeton University Press, 1971), pp. 215-234.

16. Günter Moltmann, "Steamship Transport of Emigrants from Europe to the United States, 1850-1914: Social, Commercial and Legislative Aspects," In: *Maritime Aspects of Migration,* Klaus Friedland, ed., (Köln und Wien, Bohlau Verlag, 1989), pp. 309-320.

17. Adna Ferrin Weber, *The Growth of Cities in the Nineteenth Century,* (Ithaca: Cornell University Press, 1963), p.22.

18. Josh Mason, "Would Slavery Have Ended Without the Civil War?" *U.S. History of Economics,* (April 1, 2009), p. 1.

19. Jacob Riis, *How the Other Half Lives,* (New York: Penguin Books, 1997).

20. Upton Sinclair, *The Jungle,* (New York: Barnes and Noble Classics, 2004).

21. Karl Marx, *Das Kapital: Kritik der politischen Oekonomie,* (Hamburg: Meisner, 1872).

22. Alice Wexler, *Emma Goldman: An Intimate Life,* (New York: Alfred Knopf, 1931), p. 24.

23. Ralph Tomlinson, *Population Dynamics,* (New York: Random House, 1965) p. 41.

24. Paul R. Amato, Alan Booth, David Johnson, and Stacy Rogers, *Alone Together: How Marriage in America is Changing,* (Cambridge, MA: Harvard University Press, 2007).

25. Amy Hewes, "Electrical Appliances in the Home," *Social Forces,* vol. 9, no.2, (December 1930), p. 237.

26. Frank Newport, "America's Church Attendance Inches Up in 2010," (Princeton, NJ: *Gallup Poll*), pp. 1-7.

27. David H. Wilson, "The Histiography of Science and Religion," In: Gary B. Ferngren, *Science and Religion: A Historical Introduction,* (Baltimore: Johns Hopkins University Press, 2002), 21-23.

28. www.gallup.com/pollIncreasing-Number-No-Religion-Identity.aspx.

29. Rodney Stark and Roger Finke, *Acts of Faith:Explaining the Human Side of Religion,* (Berkeley, CA: The University of California Press, 2004), p. 54.

30. N. J. Demerath III, "Secularization Extended: From Religious Myth to Cultural Commonplace," In: R. K. Fern, ed. *The Blackwell Companion to the Sociology of Religion,* (Malden, MA: Blackwell Publishers, 2001), pp. 211-228.

31. Massimo Pivetti, "Military Spending as a Burden of Growth: an Underconsumptionist Critique," *Cambridge Journal of Economics,* vol. 16, no. 4, (1992), p. 373.

32. Stuart W. Leslie, *The Cold War and American Science,* (New York: Columbia University Press, 1993), p. 73.

33. Thomas Misa, "Military Needs, Commercial Realities and the Development of the Transistor," In: Roe Smith, *Military Enterprise and Technological Change,* (Cambridge, MA: The MIT Press, 1985), p. 287.

34. No author, "What Are the Main Exports of the United States?" *Blurit,* (Norwich, Norfolk, UK, n.d.), p. 1.

35. Paul Krugman, "Increasing Returns and Economic Geography," *Journal of Political Economy,* vol. 99, (1991), p. 483.

36. Zhang Hu and Mohsin Khan, "Why Is China Growing So Fast?" *International Issues,* (1997), pp. 1-10.

37. Beata Wysocki, "Many Youths Splurge Mostly on US Goods," *Wall Street Journal,* (June 26, 1997), p. A1.

Bibliography

Books

Adler, Dennis, *Guns of the Civil War*, (Minneapolis, MN: Zenith Press, 2011).

Amato, Paul R., Alan Booth, David Johnson, and Stacy Rogers, *Alone Together: How Marriage in America is Changing*, (Cambridge, MA: Harvard University Press, 2007).

Arrington, Leonard J., *Beet Sugar in the West*, (Seattle: The University of Washington Press, 1966).

Atkins, Philip, *Dropping the Fire: The Decline and Fall of the Steam Locomotive*, (Bedfordshire, U.K.: Irwell Press, 1999).

Bain, David Howard, *Empire Express: Building the First Continental Railroad*, (New York: Viking Press, 2000).

Baker, Brian, *Masculinity in Fiction and Film: Representing Men in Popular Genres*, (London: Continuum, 2006).

Baran, Stanley J., *Introduction to Mass Communication*, (New York: McGraw-Hill, 2001).

Baratz, Morton S., *The Union and the Coal Industry*, (Westport, CT: Greenwood Press, 1983).

Bellesiles, Michael A., *Arming America: The Origin of a National Gun Culture*, (New York: Alfred Knopf, 2000).

Berger, Michael, *The Devil Wagon in God's Country: The Automobile and Social Change in Rural America 1893-1929*, (Hamden, CT: Archon Books, 1979).

Bernal, John Desmond, *Science and Industry in the Nineteenth Century*, (London, Routledge, 1953).

Bining, Arthur Cecil, *A History of the United States*, (New York: Charles Scribner's Sons, 1950).

Blood-Patterson, Peter, *Rise Up Singing: The Group Singing Song Book*, (Bethlehem, PA: Sing Out Corp., 1988).

Boyer, Carl B., *A History of Mathematics*, (New York: Wiley, 1991).

Brown, John Porter, *Turkish Evening Entertainment*, (New York: Adamant Corp., 2004).

Brown, Lyn Mikel, *Raising Their Voices: The Politics of Girls' Anger*, (Cambridge, MA, Harvard University Press, 1999).

Bruce, Robert V., *Alexander Graham Bell and the Conquest of Solitude*, (Ithaca, NY: Cornell University Press, 1990).

Budd, Lucy, "The View from the Air," In: Phillip Vanini, *The Cultures of Alternative Mobilities*, (Farnham, Surrey, UK: Ashgate, 2009).

Bularzik, Mary, "Sexual Harassment at the Workplace," IN: James Green, Editor, *Workers' Struggles, Past and Present*, (Philadelphia: Temple University Press, 1983).

Burns, James MacGregor, *The Lion and the Fox*, (New York: Harcourt, Brace and World, 1956).

Burroughs, James E., L.J. Shrum and Aric Rindfleisch, "Does television viewing promote materialism?" In: S.M. Broniarczyk and K. Nakamoto, Eds. *Advances in Consumer Research*, (Valdosta, CA: Association for Consumer Research, 2002).

Burrows, Gideon, *The No Nonsense Guide to the Arms Trade*, (London: Verso, 2002).

Caro, Robert A., The Years of Lyndon Johnson: Means of Ascent, (New York: Vintage Publishers,1991).

Ceram, C.W., *The Archaeology of the Cinema*, (London: Thomas and Hudson, 1965).

Chandler, Charlotte, *Ingrid: Ingrid Bergman, A Personal Biography*, (New York: Simon and Schuster, 2007).

Chester, Edward W., *Television and American Politics,* (New York: Sheed and Ward, 1969).

Coleman, Terry, *Going to America*, (New York: Pantheon Books, 1972).

Cook, David, *A History of Narrative Film*, (New York: Norton, 1981).

Cooper, Grace Rogers, *The Invention of the Sewing Machine*, (Washington, DC: The Smithsonian Institute, 1968).

Cross, Michael S., *The Workingman in the 19th Century*, (Toronto: Oxford University Press, 1974).

Davis, George T., *A Navy Second to None: The Development of Modern American Naval Policy*, (Westport, CT: Greenwood Press, 1971).

Demerath III, N.J., "Secularization Extended: From Religious Myth to Cultural Commonplace," IN: R.K Fern, Ed. *The Blackwell Companion to the Sociology of Religion*, (Malden, MA: Blackwell Publishers, 2001).

Devol, George H., *Forty Years a Gambler on the Mississippi*, (New York: Henry Holt, 1926).

Drache, Hugh M., *The Day of the Bonanza*, (Fargo, ND: North Dakota Institute of Regional Studies, 1964).

Dunning, John, *On the Air: The Encyclopedia of Old Time Radio*, (New York: Oxford University Press, 1998).

Durant, Will, *Caesar and Christ*, (New York: Simon and Schuster, 1944).

Edwards George C. and Stephen J. Wayne, *Presidential Leadership*, (New York: St. Martin's Press, 1985).

Egan, Timothy, *The Worst Hard Times* (Boston and New York: Houghton Mifflin Co., 2006).

Everson, George, *The Story of Television: The Life of Philo T. Farnsworth*, (New York, W.W. Norton & Co., 1949).

Falk, Ursula A. and Gerhard Falk, *Youth Culture and the Generation Gap*, (New York: Algora Publishing Co., 2005).

Fargis, Paul, ed., *The New York Public Library Desk Reference, 3rd Edition*, (New York: Macmillan Publishers, 2001).

Ferguson, Niall, *The House of Rothschild*, (New York: Penguin Books, 1999).

Finiston, Monty, ed., *Oxford Illustrated Encyclopedia of Invention and Technology*, (Oxford, England: Oxford University Press, 1992).

Fink , James Jay, *The Car Culture*, (Cambridge, MA: The MIT Press, 1975).

Fisher, Bonnie S., *Unsafe in the Ivory Tower*, (Thousand Oaks, CA: Sage, 2010).

Fleischman, Bill, Al Pearce and Junie Donleavy, *The Unauthorized Nascar Fan Guide*, (New York: Visible Ink Press, 2003).

Foucault, Michael, *The History of Sexuality: An Introduction*, (New York: Vintage-Random, 1980).

Freudenthal, Elisabeth S., *The Aviation Business from Kitty Hawk to Wall Street*, (New York: The Vanguard Press, 1940).

Gabler Neal, An *Empire of Their Own: How the Jews Invented Hollywood*, (New York: Crown Publishers, Inc., 1988).

Gabriel Richard A. and Karen S. Metz, *A Short History of War: the Evolution of Warfare and Weapons*, (Carlisle Barracks, PA: Strategic Study Institute, U.S. Army War College, 1992).

Gegas, Viktor and Monica Seff, "Families and Adolescents: A Review" In: A. Booth, ed., *Contemporary Families: Looking Forward, Looking Back*. (Minneapolis: National Council on Family Relations, 1991).

Gibbs-Smith, Charles H., *Aviaton: An Historical Survey from Its Origins To the End of World War II*, (London: Science Museum, 1970).

Gilbert, Alan, *The Mega-city in Latin America*, (New York: United Nations University Press, 2000).

Gillespie, C. C., *The Montgolfier Brothers and the Invention of Aviation, 1783-1784*, (Princeton, NJ: Princeton University Press, 1983).

Gillian, S.C., *Inventing the Ship*, (Chicago: Follett Publishing, 1935).

Goodsell, Wyllistine, *A History of Marriage and the Family*, (New York: The Macmillan Co., 1934).

Gordon, George N., *Classroom Television*, (New York: Hastings House, 1971).

Goyder, John, *The Silent Minority: Non-Respondents on Sample Surveys*, (Boulder, CO: Westview Publishing, 1987).

Graham, Philip, Showboats: *The History of an American Institution*, (Austin: University of Texas Press, 1951).

Green, Lorne E., *Chief Engineer: Life of a Nation Builder*, (Toronto: The Dundurn Press, 1993).

Grodinsky, Julius, *Transcontinental Railway Strategy, 1869-1893*, (Philadelphia: University of Pennsylvania Press, 1962).

Grosser, Morton, *Diesel: the Man & the Engine*, (New York: Athenaeum, 1978).

Guedalla, Phillip, *The Hundred Years*, (New York: Doubleday, Doran & Co. 1937).

Gutman, Herbert, "The Worker's Search for Power," In: H. Wayne Morgan, ed., *The Gilded Age*, 2nd. Ed. (Syracuse: The University of Syracuse Press, 1970),

Hansen, Miriam, *Babel and Babylon*, (Cambridge: The Harvard University Press, 1991).

Harley, C. Knick, "The Shift from Sailing Ships to Steamships, 1850-1890: A Study in Technological Change and Its Diffusion," In: *Essays on a Mature Economy*, Donald A. McCloskey, Editor (Princeton, N.J. Princeton University Press, 1971).

Harris, Abram L., "Marx on Capital and Labor," in: H.C. Harland, ed., *Readings in Economics and Politics*, (New York: Oxford University Press, 1961).

Harrison, James P., *Mastering the Sky: A History of Aviation From Ancient Times to the Present*, (New York: Saperdon, 1996).

Harrison, Mark, *The Economics of World War II*, (Cambridge, UK: Cambridge University Press, 1998).

Hawkridge, David and John Robinson, *Organizing Educational Broadcasting*, (Paris: UNESCO, 1982).

Hochschild, Arlie, *The Time Bind: When Work Becomes Home and Home Becomes Work.* (New York: Metropolitan Books, 1997).

Holbroook, Stewart, *The Story of American Railroads,* (New York: Crown, 1947).

Holliday, Jim, "A History of Modern Pornographic Film and Video," In: James Elias, ed., *Porn 101:Eroticism, Pornography and the First Amendment,* (Amherst, NY, Prometheus, 1999).

Hood, Thomas, "Song of the Shirt," in R.B. Inglis, *Adventures in English Literature,* (Toronto: W.J. Gage, 1952).

Hounshell, David A., *From the American System to Mass Production: The Development of Manufacturing in the United States,* (Baltimore, MD, The Johns Hopkins University Press, 1984).

Howe, Irving, *World of Our Fathers,* (New York: Harcourt Brace Jovanovich, 1976.

Hughes, Ivor and David Ellis Evans, *Before We Went Wireless,* (Images from the Past, Burlington, VT, 2011).

Hunter, Louis C., *Steam Boats on the Western Rivers,* (New York: Dover Publications, 1993).

Hutchins, Leigh and Amy Harrison, *A History of Factory Legislation,* (London: Routledge, 1903).

Ifrah, Georges, *The Universal History of Computing,* (New York: John Wiley and Sons, 2001).

Jackson, Kenneth T., Crabgrass *Frontier: The Suburbanization of the United States,* (New York: Oxford University Press, 1985).

Jacobs, Jane, *Death and Life of Great American Cities,* (New York: Random House, 1961).

John, Richard R., *Network Nation: Inventing American Telecommunications,* (Cambridge, MA: Harvard University Press, 2010).

Kabir, Nareen M., *Bollywood: The Indian Cinema Story,* (London: Channel 4 Books, 2001).

Kelly , Ralph, *Matthias W. Baldwin: Locomotive Pioneer,* (New York: The Newcomen Society, 1946).

Kelly, Fred C., *The Wright Brothers,* (New York: Harcourt, Brace and Co., 1945).

Kidwell, Claudia and Margaret C. Christman, *Suiting Everyone: The Democratization of Clothing in America,* (Washington DC: Smithsonian Institution Press, 1974).

Kimmel, Michael S., "Masculinity as Homophobia," In: Paul Murphy et. al., *Feminism and Masculinities,* (New York: Oxford University Press, 2004).

King, Henry C. A., *History of the Telescope,* (New York: Dover Books, 2003).

Klein, Maury, *The Life and Legend of Jay Gould,* (Baltimore, MD: Johns Hopkins University Press, 1986).

Kline, Ronald R., *Consumers in the Country: Technology and Social Change in Rural America,* (Baltimore: Johns Hopkins University Press, 2000).

Landes, David, *Revolution in Time,* (Cambridge, MA: The Harvard University Press, 1983).

Law, Rodney J., *James Watt and the Separate Condenser,* (London: Science Museum Monograph 46, 1969).

Leishman, J. Gordon, *Principles of Helicopter Aerodynamics,* (New York: Cambridge University Press, 2006).

Leslie, Stuart W., *The Cold War and American Science,* (New York: Columbia University Press, 1993).

Levine, Louis, *The Women's Garment Workers: A History of the International Ladies Garment Workers Union,* (New York: B.W. Huebsch, 1924).

Lloyd, James T., *Lloyd's Steam Boat Directory and Disasters on the Western Waters*, (Philadelphia: Jasper Harding, Publishers, 1856).

Lubrano, Anteresa, *The Telegraph: How Technology Innovation Caused Social Change*, (New York: Garland Publishing Inc., 1977).

Lynd, Robert S. and Helen Merrill Lynd, *Middletown: A Study in Contemporary American Culture*, (New York: Harcourt, Brace and Co., 1929).

MacDonald, Philip B., *The Saga of the Sea: The Story of Cyrus W. Field and the Laying of the First Atlantic Cable*, (New York: Wilson-Erickson, Inc. 1937).

Mackay, James A., *Airmails 1870-1970*, (London: B. T. Batesford, 1971).

Maddison, Angus, *The World Economy: A Millenial Perspective*, (Paris: OECD, 2001).

Manger, Louis N., *A History of the Life Sciences*, (New York: M. Dekker, Publishers, 2002).

Markowitz, Gerald and David Rosner, *Slaves of the Depression: Workers' Letters About Life on the Job*, (Ithaca, Cornell University Press, 1987).

Marx, Karl, *Das Kapital: Kritik der politischen Oekonomie*, (Hamburg: Meisner, 1872).

Mast, Gerald, *A Short History of the Movies*, (Chicago: The University of Chicago Press, 1981).

McGrane, Reginald C., *The Panic of 1837*, (New York: Russell and Russell, 1965).

Medved, Michael, *Hollywood vs. America: Popular Culture and the War on Traditional Values*, (New York: Harper Collins, 1992).

Moltmann, Günter, "Steamship Transport of Emigrants from Europe to the United States 1850-1914" IN: Klaus Friedland, ed., *Maritime Aspects of Migration*, (Cologne: Bohlau Verlag, 1989).

Morris, Jenny, *Women Workers and the Sweated Trades: The Origins of Minimum Wage Legislation*, (Aldershot, England: Hants Publishing, 1986).

Nasaw, David, *Going Out: The Rise and Fall of Public Amusements*, (Cambridge: The Harvard University Press, 1993).

Neumann, George C., *The History of Weapons of the Revolutionary War*, (New York: Harper and Row, 1967).

Nickles, David Paul, *Under the Wire: How the Telegraph Changes Diplomacy*, (Cambridge, MA: Harvard University Press, 2003).

O'Malley, Michael, *Keeping Watch: A History of American Time*, (Washington: Smithsonian Institution Press, 1990).

Olsen, Laurie, *Made in America: Immigrant Students in Our Public Schools*, (New York: New Press, 1997).

Orleck, Annelise, *Common Sense and a Little Fire: Women's Working Class Politics in the United States*, (Chapel Hill, The University of North Carolina Press, 1995).

Orum, Anthony M., *The Centrality of Place: The Urban Imagination*, (Ann Arbor: The University of Michigan, 2004).

Pauly, Philip J., *Fruits and Plains: The Horticultural Transformation of America*, (Cambridge, MA: The Harvard University Press, 2008).

Pessen, Edward, *Most Uncommon Jacksonians*, (Albany, NY: State University of New York Press, 1967).

Pierce, Benjamin, *Types and Programming Languages*, (Cambridge, MA: The MIT Press, 2004).

Poincare, Lucien, *The New Physics and Its Evolution*, (London: Paul, Trench & Co., 1907).

Porter, Michael, *The Competitive Advantage of Nations*, (New York: The Free Press, 1990).

Potter, George W., *To the Golden Door: The Story of the Irish in Ireland and America*, (Boston: Little Brown, & Co., 1960).

Poundstone, William, *Fortune's Formula*, (New York: Hill and Wang, 2005).

Poutanen, Mary Ann, *For the Benefit of the Master, 1820-1842*, (Montreal: McGill University, 1985).

Quinn, Dermott, *The Irish in New Jersey*, (New Brunswick, NJ: Rutgers University, 2004).

Rieff, Lynn Anderson, *Rousing the People of the Land*, (Auburn, AL: Auburn University, 1995).

Riis, Jacob, *How the Other Half Lives*, (New York: Penguin Books, 1997).

Robertson, Ross M., *History of the American Economy*, (New York: Harcourt, Brace and Co., 1955).

Rolt, L.T C., *Thomas Newcomen: The Pre-history of the Steam Engine*, (London: Newton Abbott, 1963).

Ross, Robert J.S., *Slaves to Fashion: Poverty and Abuse in the Sweat Shops*, (Ann Arbor: The University of Michigan Press, 2004).

Ross, Steven J., *Working Class Hollywood: Silent Films and the Shaping of Class in America*, (Princeton, NJ: Princeton University Press, 1998).

Rowsome, Jr., Frank, *The Verse by the Side of the Road: the Story of the Burma Shave Signs*, (Brattleboro, VT: S. Greene Press, 1965).

Scharff, Virginia, *Taking the Wheel: Women and the Coming of the Motor Age*, (New York: Free Press, 1991).

Schatzkin, Paul, *The Boy Who Invented Television*, (Silver Springs, MD: Tamcom Books, 2002).

Schmiechen, James, *Seated Industries and Sweated Labor*, (Urbana, Illinois: University of Illinois Press, 1984).

Schwarzlose, Richard, *Nation's Newsbrokers: Vol. 1: The Formative Years* (Evanston, IL: Northwestern University Press, 1989).

Scott, Joan Wallach, *Gender and the Politics of History*, (New York: Columbia University Press, 1988).

Sharp, Lauriston, "Steel Axes for Stone Age Australians," IN: Edward H. Spice, *Human Problems in Technological Change*, (New York: The Russell Sage Foundation, 1952).

Shepherd, Donald and Robert Slatzer, *Duke: The Life and Times of John Wayne*, (New York: Kensington Publishers, 1985).

Sherry, J. and T. Salvador, "Running and Grimacing: The Struggle for Balance in Mobile Work," In: Barry Brown, ed., *Wireless World: Social and Interactional Aspects of the Mobile Age*, (London: Springer Verlag, 2006)

Shulman, Seth, *The Telephone Gambit:Chasing Alexander Graham Bell's Secret*, (New York: W.W. Norton, Inc., 1988).

Sinclair, Upton, *The Brass Check: A Study of American Journalism*, (Urbana, IL: The University of Illinois Press, 2003).

Sinclair, Upton, The Jungle, (New York: Modern Library, 2002).

Sklar, Robert, "Oh Althusser! Histiography and the Rise of Cinema Studies," In: Robert Sklar and Charles Musser, eds., *Resisting Images: Essays on Cinema and History*, (Philadelphia: Temple University Press, 1999).

Skoric, Sophia and George Vid Tomashevich, *Serbs in Ontario: A Socio-cultural Description*, (Toronto: Serbian Heritage Academy, 1974).

Sloan, Kay, *The Loud Silents: Origins of the Social Problem Film*, (Urbana, IL: The University of Illinois Press, 1988).

Smith, Edgar C., A *Short History of Naval and Marine Engineering*, (Cambridge: University Press, 1938).

Sontag, Susan, "The Double Standard of Aging," In: Juanita H. Williams, ed. *Psychology of Women*, (New York: Academic Press, 1970).

Sparling, Tobin A., *The Great Exhibition: A Question of Taste*, (New Haven: Yale Center for British Art, 1983).

Standage, Tom, *The Remarkable Story of Telegraphy and the Nineteenth Century On Line Pioneers*, (New York: Walker Publishing Co., 1999).

Stansell, Christine, *City of Women: Sex and Class in New York*, 1789-1860. (Urbana, IL: The University of Illinois Press, 1987).

Stark, Rodney and Roger Finke, *Acts of Faith: Explaining the Human Side of Religion*, (Berkeley, CA: The University of California Press, 2004).

Steedman, Mercedes, "Skill and Gender in the Canadian Clothing Industry, 1890-1940" in: Craig Herron and Robert Storey, eds. *On the Job: Confronting the Labor Process in Canada*, (Montreal and Kingston: McGill and Queen's University Press, 1986).

Stein, Leon, *Out of the Sweatshop: The Struggle for Industrial Democracy*, (New York: Quadrangle/New Times Book Co., 1977).

Stone, Melville E., "Fifty Years a Journalist," *Proceedings of the First National Newspaper Conference*, (Madison, Wisconsin: The University, 1913).

Stover, John F., *The Life and Decline of the American Railroad*, (New York: Oxford University Press, 1970).

Strachan, Hew, *The Illustrated History of the First World War*, (New York: The Oxford University Press, 2000)

Strasser, Susan, *Never Done*, (New York: Henry Holt & Co., 2000).

Susskind, Charles, "Armstrong, Edward Howard," *Dictionary of Scientific Biography*, (New York: Charles Scribner and Sons, 1970).

Taveau, Augustin L., "Modern Farming in America," (Washington DC: *Report for the Committee on Agriculture for 1874).*

Teaford, Jon C., The *Metropolitan Revolution: The Rise of Post-Urban America*, (New York: Columbia University Press, 2006).

Thomas, William I., *The Child in America*, (New York: Alfred Knopf, 1929).

Thompson, Edward P., *The Making of the English Working Class*, (New York: Vintage Books, 1966).

Thompson, Lana, "Menarche," In: *International Encyclopedia of Marriage and the Family* (New York:, Macmillan Publishers, 2003).

Thompson, Robert L., *Wiring a Continent: The History of the Telegraph Industry in the United States*, 1832-1836, (Princeton, NJ: The Princeton University Press, 1947).

Tomlinson, Ralph, *Population Dynamics*, (New York: Random House, 1965).

Trachtenberg, Alan, *The Incorporation of America*, (New York: Hall and Wang, 1982).

Tropiano, Stephen, *Obscene, Indecent, Immoral and Offensive: One Hundred Years of Censored, Banned and Controversial Films*, (New York: Limelight Editions, 2009).

Tuchman, Barbara, *The Guns of August*, (New York: Dell Publishing Co., 1963).

Twain, Mark, *Life on the Mississippi*, (New York: Signet Classics, 2009, original 1883).

U. S. Department of Labor, Bureau of Labor Statistics, *Occupational Outlook Handbook 2010-2011*, (Washington, DC: U.S. Government Printing Office, 2010).

Urwin, George J.W., *The United States Infantry*, (New York: Sterling Publishing Co., 1991).

Uviller, H. Richard and William G. Merkel, *The Militia and the Right to Arms or How the Second Amendment Fell Silent*, (Durham. NC: Duke University Press, 2002),

Veblen, Thorstein, *The Theory of the Leisure Class: An Economic Study of Institutions*, (New York: Macmillan, 1902).

Waldinger, Roger D., *Through the Eyes of the Needle: Immigrants and Enterprise in New York Garment Trades*, (New York: New York University Press, 1986).

Warwick, Kevin I., *Cyborg*, (Urbana: The University of Illinois Press, 2004).

Wattenberg, Ben J. and Richard A. Scammon, *An Unexpected Family Portrait of 194,067,296 Americans*, (Garden City, NY: Doubleday, 1965).

Weber, Adna Ferrin, *The Growth of Cities in the Nineteenth Century*, (Ithaca: Cornell University Press, 1963).

Wernick, Andrew, *Promotional Culture: Advertising, Ideology and Symbolic Expression*, (London: Sage, 1992).

Wexler, Alice, *Emma Goldman: An Intimate Life*, (New York: Alfred Knopf, 1931).

White, John H., "Grice and Long: Steam Car Builder," in: Jack Salzman, ed. *Prospects: An Annual of American Cultural Studies*, (New York: Burt Franklin & Co., 1976).

Wik, Reynold M., *Henry Ford and Grass Roots America*, (Ann Arbor: The University of Michigan Press, 1972).

Wik, Reynold M., Steam Power on the American Farm, (Philadelphia: The University of Pennsylvania Press, 1953).

Williams, Linda, *Hard Core: Power, Pleasure, and the 'Frenzy of the Visible'*, (Berkeley: The University of California Press, 1999).

Wilson, David H., "The Histiography of Science and Religion," In: Gary B. Ferngren, Science and Religion: A Historical Introduction, (Baltimore: Johns Hopkins University Press, 2002).

Wollstein, Harold J., *Strangers in Hollywood*, (London: The Scarecrow Press, 1994).

Yates, JoAnne and Robert J. Benjamin, "The Past and Present as a Window on the Future," In: Michael S. Scott Morrison, ed., *The Corporation of the 1990's: Information Technology and Organizational Transformation*, (New York: Oxford University Press, 1991).

Journals

Adam, Marcel, Timothy A. Keller and Jacquelin Cynkar, "A Decrease in Brain Activation Associated with Driving When Listening to Someone Speak," *Brain Research*, vol. 1205 (2008).

Adams, Clark E., Robert D. Brown, Billy J. Higginbotham, "Developing a Strategy Plan for Future Hunting," *Wildlife Society Bulletin*, vol. 32, no.4 (Winter 2004).

Alexander, Joseph H., "Tracked Landing Vehicles, *World War II* , vol. 13, no. 4 (November 1998).

Alexander, Mary and Marylin Childress, "The Zimmerman Telegram," *Social Education*, vol. 45, no. 4, (April 1981).

Aluise, Susan J., "Reasons Airline Earnings Will Tank in 2012," *Investment Place Media*, (October 10, 2011).

Anderson, Craig A. and Karen E. Dill, "Video Games and Aggressive Thoughts, Feelings, and Behavior in the Laboratory and in Life," *Journal of Personality and Social Psychology*, vol. 78, no.4, (2000).

Anderson, Craig, "Effect of Violent Movies and Trait Hostility on Hostile Feelings and Aggressive Thoughts," *Aggressive Behavior*, vol. 23, (1997).

Anderson, Virginia D., "Thomas Minor's World: Agrarian Life in 17th Century New England," *Agricultural History*, vol.82 (Fall 2008).

Bacon, Elizabeth M., "Marketing Sewing Machines in the Post Civil-War Years," *Bulletin of the Business Historical Society*, vol. 20, no.3, (June 1946):90-91.

Bainer, Roy, "Preliminary Trials of a New Type Mower," *Agricultural Engineering*, vol. 12, (May 1931).

Bainer, Roy, "Harvesting and Drying Rough Rice in California," *Rice Journal*, vol. 47, (July 1944).

Ballowe, James, "Thoreau and Etzler: Alternative Views of Economic Reform," *Midcontinent American Studies Journal*, vol. 11, (1970).

Bardwick, John and S. Schuman, "Portrait of American Men and Women in TV Commercials," *Psychology*, vol. 4, no.4, (1967).

Barth, B.J., "Television Used in the Teaching of Science," *School Science and Mathematics*, vol. 58, (1958).

Barth, Matthew and Kanok Boriboonsomsin, "Traffic Congestion and Greenhouse Gases," *Access*, vol. 35, (Fall, 2009).

Bassoli, Arriana, et. al., "Underground Aesthetics: Rethinking Urban Computing," *Pervasive Computing*, vol. 6, (2007).

Bazzini, Doris G., "The Aging Woman in Popular Film," *Sex Roles*, vol. 36:531-543.

Becker, Henry J. and Jay Ravitz, "The Influence of Computer and Internet Use on Teachers' Pedagogical Practices and Perceptions," *Journal of Research and Computing in Education*, vol.31, no.4, (1999).

Beisenherz, Paul C., "A Comparison of the Quality and Sequence of Television and Classroom Science," *Journal of Research in Science Teaching*, vol.10, (1973).

Berkey, Catherine S. Jane D. Gardner, A. Lindsay Frazier and Graham A. Colditz, "Relation of Childhood Diet and Body Size to Menarche and Adolescent Growth in Girls," *American Journal of Epidemiology*, vol. 152, no.5, (2000).

Berkowitz, Justin, "A History of Car Radios," *Car and Driver*, (October 2010).

Berkowitz, Len, and J. T. Alioto, "The Meaning of an Observed Event as a Determinant of its Aggressive Consequences," *Journal of Personality and Social Psychology*, vol. 28, (1973).

Bilstein, Roger E., "The First Air War 1914-1918," *The Journal of American History*, vol. 78, no.4, (March 1992).

Birbeck, Christopher, "Controlling New Mexico Juveniles' Possession and Use of Firearms," *Justice Research and Policy*, vol. 1, (1999).

Bisignanni, Giovanni, "Airlines," *Foreign Policy*, (January/February 2006).

Bjerga, Alan, "U.S. Farm Exports May Increase to Record $137 Billion," *Bloomberg*, (May 26, 2010).

Boles, Donald and Gary Rupnow, "Local Governmental Functions Affected by the Growth of Corporate Agricultural Land Ownership," *Western Political Quarterly*, vol. 32, (1979).

Bowen, John, "Network Changes, Deregulation, and Access in the Global Airline Industry," *Economic Geography*, vol. 74, no.4, (2002).

Brevik, Eric C. and Alfred Hartemink, "Early Soil Knowledge and the Birth and Development of Soil Science," *Catena*, vol. 83 (2010).

Breyer, Steven, "Anti-deregulation Revisited," *Business Week*, (January 20, 2011):

Brown, J. D., "Sexy Media Matter: Exposure to Sexual Content in Music, Movies, Television and Magazines," *Pediatrics*, vol.117, no.4, (2006).

Burch, William R., "Resources and Social Structure: Some Conditions of Stability and Change," *The Annals*, vol. 389, (1973).

Cailliau, Robert, "A Short History of the Web," *Net Valley*, (November 2, 1995).

Cain, Bruce E., "The Constituency Service Basis of the Personal Vote for the U.S. Representatives and British Members of Parliament," *American Political Science Review*, vol. 78, no.1, 1984).

Carlson, Edward and R. Wald, "Einstein's Universe," *Aviation and Space*, vol. 6, (1979).

Carlson, John B. ""Lodestone Compass: Chinese or Olmec Primacy?" *Science*, vol. 4205, no. 189, (September, 1975).

Carroll, Noel, "The Power of Movies," *Daedalus*, v. 114, no.4, (Fall 1985).

Chapman, Samuel, "Blaise Pascal Tercentenary of the Calculating Machine," *Nature*, vol. 150, (October 31, 1942).

Christensen, Tricia Ellis, "How has the average age at marriage changed over time?" *Wise Geek*, (March 26, 2011).

Clarke, Deborah, "Eudora Welty's Losing Battles: Cars and Family Values," *Mississippi Quarterly*, vol. 62, no. 13, (April 2009).

Clough, Alfred, "Nervous Strain Due to Automobile Driving," *Horseless Age*, (September 23, 1903).

Cook, Philip, J. S. Moliconi, and T.B. Cole, "Regulating Gun Markets," *Journal of Criminal Law and Criminology*, vol. 86, no.1, (1995).

Cox, Matthew, "Troops in Iraq give Thumbs Up to Land Warrior," *Army Times*, (June 23, 2007).

Craig, Steven, "How America Adopted Radio," *Journal of Broadcasting and Electronic Media*, (June 2004).

Dailey, Janelle, "Sewing is Relaxing and Productive," *Countryside and Small Stock*, (June 2005).

Dexter, Robert and Aaron Bernstein, "Inside a Chinese Sweatshop: A Life of Fines and Beatings," *Bloomberg Business Week*, (October 2, 2000).

DiCamillo Jodi and Joseph M. Schaefer, "Internet Program Impacts Youth Interest in Hunting," *Wildlife Society Bulletin*, vol.28, no.4, (2000).

Diffie, Whitfield and Susan Landau, "Internet Eavesdropping: A Brave New World of Wiretapping," *Scientific American*, (September 2008).

Dillman, Don A., "Reducing Refusal Rates for Telephone Interviews," *Public Opinion Quarterly*, vol. 40, (1976).

Dunning, David, "The Average Height of Humans Over Time," *Health*, (September 10, 2010).

Editorial, "Contempt of Court or Contempt for a Vicious System? Jail Sentences on Frail Women," *The Ladies' Garment Worker*, (January 1912).

Edwards, Mark, "The Sunny Old World of Sex on Wheels," *Daily Mail*, (February 15, 1992).

Eisenhower, Milton, "Uncle Sam Chats with his Dairymen," *Radio Age: The Magazine of the Hour*, (October 1926).

Ennels, Jerome, ""The Wright Stuff: Pilot Training at America's First Civilian Flying School," *Air Power History*, vol. 49, no. 4, (Winter 2002).

Ferguson, Christopher J., "Violent Video Games and Aggression," *Criminal Justice and Behavior*, vol. 35, no.3, (March 2008).

Field, Alexander, "The Telegraphic Transmission of Financial Asset Orders," *Research and Economic History*, vol. 18, (1998).

Field, Alexander J. "The Magnetic Telegraph, Price and Quantity Data and the New Management of Capital," *The Journal of Economic History*, vol. 52, no.2, (June 1992).

Figer, Jessica, "Twilight of the Sewing Machine Repairman," *Metropolis*, (February 11, 2011).

Fineman, Stephen, "Getting the Measure of Emotion - and the Cautionary Tale of Emotional Intelligence," *Human Relations*, vol.57, no.6, (2004)..

Fisher, Samuel, "Identifying Video Game Addiction in Children and Adolescents," *Addictive Behavior*, vol. 19, (1994).

Flanagan, Robert, "How Western Union Went from Bad to Worse," *Business Week* (January 28, 1984).

Flegel K.M., M.D. Carroll, G. L. Ogden and L. R. Curtin, "Prevalence and Trend in Obesity Among US Adults," *Journal of the American Medical Association*, vol. 303, no. 3, (2010).

Fowler, John, "On Cultivating by Steam," *Journal of the Society of Arts*, vol. 4, (February 1856).

Franke, Katherine, "What's Wrong With Sexual Harassment?" *Stanford Law Review*, vol. 49, (1997).

Funk, Jeanne B., G. Flores and D.D. Buchman, "Rating Electronic Games: Violence is in the Eye of the Beholder," *Youth and Society*, vol. 30, (1999).

Furfey, Paul Hanley, "Steam Power: A Study in the Sociology of Invention," *The American Catholic Sociological Review*, vol. 5, no.3, (October 1944).

Gibbs, Adrienne Samuels, "Make Working from Home Work for You," Ebony, vol. LXV no. 12, (October 2010).

Goen, C. C., "Jonathan Edwards: A New Departure in Eschatology," *Church History*, vol. 28, (1959).

Goldschmidt, Walter, "Large Scale Farming and the Rural Social Structure," *Rural Sociology*, vol. 43, (1978).

Goldstein, Edward, "Wings of Gold: One Hundred Years of U.S. Navy Air Power," Aerospace America, (September 2011).

Gorman, Peter, "Torture by Taser," *Fort Worth Weekly*, (June 23 2005).

Gough, David and Don Elbourne, "Systematic Research Synthesis to Inform Policy, Practice and Democratic Debate," *Social Policy and Society*, no.1, (2002).

Grady, Lee, "McCormick Reapers at 100: Marketing the Machines that Revolutionized World Agriculture," *The Wisconsin Magazine of History*, vol. 84., no.3, (Spring 2001).

Griffiths, Mark D., "Videogame Addiction," *International Journal of Mental Health and Addiction*, vol. 6, (2008).

Hachman, Mark, "Facebook Now Totals 901 Million Users," PC Magazine, (August 23, 2012).

Halbrook, Stephen, "Nazi Firearms Law and the Disarming of the German Jews," *Arizona Journal of International and Comparative Law*, vol. 17, no.3 (2000).

Hale, Sarah, "The Seamstess," Godey's Lady's Book, vol 74, 1867.

Hall, Burton, "I.L.G.W.U.: Its Enemies and Its Friends," New Politics, (Fall, 1976):46-50.

Harkin, James, "Life Lines," *New Statesman*, vol. 17 (September 15, 2003).

Hawkins, Albert, "Centennial of the Covered Wagon," *Oregon Historical Quarterly*, vol. 31, no.2, (June 1930).

Hayner, Norman S.,"The Tourist Family," *Social Forces*, vol.11, no.1, (October 1932).

Hemenway, David, " The U.S. Gun Stock," *Injury Prevention*, vol. 13, (2006).

Hewes, Amy, "Electrical Appliances in the Home," *Social Forces*, vol. 9, no.2, (December 1930).

Higgins, F. H., "The Combine Parade," *The Farm Quarterly*, no.4, (Summer, 1949).

Higgins, F. Hal, "97 Years of Combining in California," *California Farmer*, vol. 25, (March 1950).

Hu, Zhang and Mohsin Khan, "Why Is China Growing So Fast?" *International Issues*, (1997).

Hugill, Peter J., "Good roads and the Automobile in the United States 1880-1929," *Geographical Review*, vol. 72, no.3, (July 1982).

Irvin, Will, "The United Press," *Harper's Weekly*, (March 28, 1914).

Jackson, Donald, "The Public Image of Lewis and Clark," *The Pacific Northwest Quarterly*, vol. 57, no. 1, (January 1966).

Jackson-Smith, Douglas and Gilbert W. Gillespie, "Impact of Farm Structural Change on Farmers' Social Ties," *Society and Natural Resources*, vol. 18, (2005).

James, Clive, "Global Status of Commercialized Transgenic Crops," *International Service for the Acquisition of Agri-biotech Applications*, Brief 42, (2010).

Johnson, James D., J.T. Jackson and L. Gato, "Violent Attitudes and Deferred Academic Aspirations: Deleterious Effects of Exposure to Rap Music," *Basic and Applied Social Psychology*, vol.16, no.1 (1995).

Kearns, Mark, "State of the Adult Film Industry," *AVN Magazine*, (March 2007).

Kessler-Harris, Alice, "Organizing the Unorganizable: Three Jewish Women and Their Union," *Labor History*, vol. 17, (Winter 1976).

Kessler-Harris, Alice, "Where are the Organized Women Workers?" *Feminist Studies*, vol 3, (Fall 1975).

Kirkpatrick, Howard J., "A controlled Archery Deer Hunt in a Residential Community," *Wildlife Society Bulletin,* vol.27, (1999).

Koch, Kathy, "Flexible Work Arrangements: Do They Really Improve Productivity?" *CQ Researcher*, vol.8, no. 30, (1998).

Krause, Kenneth R. and Leonard R. Kyle, "Economic Factors Underlying the Incidence of Large Farming Units," *American Journal of Agricultural Economics*, Vol. 42 (December 1970).

Krugman, Paul, "Increasing Returns and Economic Geography," *Journal of Political Economy*, vol. 99, (1991).

Kunze, Joel P., "A Tractor or an Automobile? A 1920's Farm Family Faces a Decision," *Agricultural and Rural Life*, vol. 5, no.3, (Winter 1991).

Kurz, Demie, "Caring for Teenage Children," *Journal of Family Issues*, vol. 23, no.6. (2002).

LaFeber, Walter, "A Note on the 'Mercantalistic Imperialism' of Alfred Thayer Mahan," *The Mississippi Valley Historical Review*, vol. 48, no. 4, (March 1962).

Lane, Frederic C., "The Economic Meaning of the Invention of the Compass," *The American Historical Review*, vol. 68, no. 3, (April 1963).

Lanford, Audri and Jim, "Home Based Business Scams," *Internet Scams*, no.61, (June 2012).

Langsford, Walter,, "What the Motor Vehicle is Doing for the Farm," *Scientific American*, vol.102, (January 15, 1910).

Lehman, Sally, "Dolly's Creator Moves Away from Cloning and Embryonic Stem Cells," *Scientific American*, (August 2008).

Leibenluft, Jacob, "Six Thousand Gallons of Regular, Please," *Slate Magazine*, (April 28, 2008).

Levy, Glen, "War of the Worlds: Top Ten Hoaxes," *Time*, (March 16, 2010).

Lipsig, Carla "Organizing Women in the Clothing Trades," *Studies in Political Economy*, vol. 22, (1987).

Longley, Robert, "Americans Are Getting Taller, Bigger, Fatter, Says CDC," *About.com U.S. Government Info.* (October 28, 2004).

Lovdal, Lynn T., "Sex Role Messages in Television Commercials: an Update," *Sex Roles*, vol. 21, (November 1989).

Luke, Carmen, "Buzzing Down the On-ramp of the Super-highway," *Social Activities*, vol. 20, no. 1, (January 2001).

Luke, Carmen, "What next? Toddlery Netizens, Playstation Thumb, Techno-literacies," *Contemporary Issues in Early Childhood*, vol.1, no.1, (1999).

Maas, Bertram, "911 Assistance Emergency Call Systems," *George Mason University Law Review*, vol 8, no.1, (1985).

Malamuth, Neil M., "Pornography, Individual Differences in Risk, etc." *Sex Roles*, vol. 66, (2012).

Marek, Jiri, "Marxism as Product of the Age of the Steam Engine," *Studies in Soviet Thought*, vol. 32, no.2, (August 1986).

Mason, Josh, "Would Slavery Have Ended Without the Civil War?" *U.S. History of Economics*, (April 1, 2009).

Mayfield, Betty, "Gerbert d'Aurillac and the March of Spain: A Convergence of Cultures," *The Mathematical Sciences Digest*, (no date).

McCullough, David, "Reversal of Fortune," *Smithsonian*, (September 2011)

McGanney, Daniel J., Fredrick B. Whitman and David A. Hill, "Western Railroads," *The Analyst's Journal*, vol. 8, no.4, (August 1952).

McIntosh, Robert, "Sweated Labor: Female Needleworkers in Industrialized Canada," *Labour*, (Fall, 1993).

McLaughlin, George F., "The Story of the Airplane Engine," *The Annals of the American Academy of Political and Social Sciences*, vol. 131, (May 1927).

McLuhan, Eric, "Marshall McLuhan Forsees the Global Village," *Marshall McLuhan Studies*, vol.1, no.4,(May 16, 2000).

Megonnell, William J. and Howard W. Chapman, "Sanitation of Domestic Airlines," *Public Health Reports*, vol. 71, no.4, (April 1956).

Metha, Suketu, "Welcome to Bollywood," National Geographic, vol. 207, no. 2, (February 2005).

Milkman, Ruth, "Redefining Women's Work: The Sexual Division of Labor in the Auto Industry During World War II," *Feminist Studies*, v. 8, (summer 1982).

Millard, Max, "Lee DeForest: Class of 1893, Father of the Electronics Age," *Northfield Mount Hermon Alumni Magazine*, (October 1993).

Monahan, A.C., "Bell Rings Down a Century," *Science News Letter*, (March 1, 1947).

Montrie, Chad, "Men Alone Cannot Settle a Country: Domesticating Nature in the Kansas –Nebraska Grassland," *Great Plains Quarterly*, vol.25, no.4, (Fall 2005).

Morehead, F. G., "Automobile vs. Country Church," *Technical World Magazine*, (vol. 18, November 1912).

Mottram, Ralph, "The Great Northern Film Company," *Film History*, vol.2, no.l, (1988).

Murphy, Alice and Peter Murphy, "Philip Parmalee: the World's First Commercial Pilot," *Michigan History Magazine*, vol. 95, no. 1, (January-February, 2011).

Musciamo, Walter, "German Condor Legion's Tactical Air Power," *Aviation History*, (September 2004).

Mustaine, Elizabeth E. and Richard Tewksbury, "Predicting Risks of Larceny Theft Victimization," *Criminology*, vol. 36, (1998).

Nash Information Service, *The Numbers, U.S. Movie Market Summary 1995-2011*, (2011).

Nava, Mica, "Consumerism Reconsidered: Busing and Power," *Cultural Studies*, vol. 51, no. 57, 1991).

Ng, B. D., and Peter Wiemer-Hastings, "Addiction to the Internet and Online Gaming," *CyberPsychology and Behavior*, vol.8, no.2, (2005).

No author, "Alan Turing," *Time Magazine*, vol. 154, no. 27 (December 31, 1999).

No author, "Amphibious Landing Vehicles of the Second World War," *Car News*, (June 11, 2007).

No author, "The Automobile in War," *Lotus Magazine*, vol.7., no.2, (November 1915).

No author, "Beauty, Cosmetic and Fragrance Stores in the US: Market Research Report" *IBISWorld*, (March 2012).

No author, "Civil Aeronautics Board Policy: An Evaluation," *The Yale Law Journal*, vol. 57, no. 6, (April 1948).

No author, "Flight Training on Ground," *Science News Letter*, vol. 53, no. 21, (May 22, 1948).

No author, "History of American Agriculture, 1776-1990" *Farmers and the Land*, (New York: The New York Times Company).

No author, "Meet QueerPeople," Motor Age, (July 14, 1904).

No author, "Pornography Statistics," *Family Safe Media*, (2012).

No author, "Squad Automatic Weapon (SAW), M249 Light Machine Gun," *Military Analysis Network*, (January 20, 1999).

No author, "Telephone," *Think Quest*, (New York: Educational Foundation, 2010).

No author, "The Automobile Age," *The Wilson Quarterly*, vol. 10, no. 5, (Winter1986).

No author, "The Old Issues Again," *Sewing Machine Times*, (December 25, 1902).

No author, "The Telegraph in America," *The Illustrated Magazine of Art*, no.6 (1853).

No author, "The Telephone in Retail Business," *Printer's Ink*, vol. 61, (November 27, 1907).

No author, "This Day in History," *http://www.history.com/this-day-in-history*

No author, "Western Electric Getting Farmers to Install phones," *Printers Ink*, vol.76, (July 27, 1911).

No author, "What Are the Main Exports of the United States?" *Blurit*, (Norwich, Norfolk, UK, n.d.).

O'Donnell, Edward T., "Cyrus McCormick Invents the Reaper," *Irish Echo*, (June 19, 2002):1.

Olsen, Michael L., "Corporate Farming in the Northwest: the Puget's Sound Agricultural Company," *Idaho Yesterdays*, vol. XIV, (Spring 1970).

Osaki, Amy Boyle, "Truly Feminine Employment," *Winterthur Portfolio*, vol. 23, no.4, (Winter, 1988).

Otswald, William, "The Modern Theory of Energetics," *The Monist*, vol. 17, (1907).

Paolozzi, Remi, "The Cradle of Motor Sport,"*Autosport*, (May 28, 2003).

Parolin B. and A. Harrington, "Changing Long Distance Passenger Markets in a Deregulated Environment," *Transportation Research Record*, vol. 1341, (1992).

Pasqaurelli, Adrianne, "Garment Workers Protest Sweat Shops in Long Island City," *Workforce Management*, (August 12, 2009).

Patterson, Ethel Lloyd, "The Cost of Being Well Dressed." *Ladies Home Journal*, Vol. 40, no.3, (April 1923).

Paxson, Frederick L., "The Railroads of the Old Northwest," *Transactions of the Wisconsin Academy of Sciences*, vol. 17, no.2, (1911).

Philippe Aries, "The Family and the City," *Daedalus*, v. 106, (1977).

Pirie, Gordon, "Cultural Crossings," *Journal of Transportation History*, vol. 29, no.1 (2008).

Pivetti, Massimo, "Military Spending as a Burden of Growth: an Underconsumptionist Critique," *Cambridge Journal of Economics*, vol. 16, no.4, (1992).

Plante, Trevor K., "New Glory to Its Already Gallant Record," *Prologue*, vol. 30, no.1, (Spring 1998).

Plowman, Lydia Joanna McPake and Chistine Stephen, "Just Picking it Up: Young children Learning with Technology at Home," *Cambridge Journal of Education*, vol. 38, no. 31, (2008).

Powers ,John B., "The Development of a New Sugar Beet Harvester," *Agricultural Engineering*, vol.29, (August 1948).

Prior, Markus, "The Incumbent in the Living Room," *The Journal of Politics*, vol. 69, no. 3, (August 2006).

Rawles, James Wesley, "Round Capacity Magazine," *The M14/M1A Magazine*, (November 17, 2008)..

Reyns, Bradford W., "A Situational Crime Prevention Approach to Cyberstalking Victimization," *Crime Prevention and Community Safety*, vol.12, (2010).

Richards, Rebekah, "Gender Construction in a Ford F-150 Commercial," *Anthropology*, November 7, 2009).

Richmond, Samuel A., "Why Are the Airlines in Trouble?" *Challenge*, vol. 10, no 8. (May 1921).

Ridge, Martin, "The Life of an Idea: Significance of Frederick Jackson Turner's Frontier Thesis," *Montana: The Magazine of Western History*, vol.41, no. 4, (Winter 1991).

Roberts, Glyn, "The Hitler of Hollywood," *Film Weekly*, (August 31, 1934).

Robinson, Eric, "The International Exchange of Men and Machines, *Business History*, vol 1, (1958).

Rosenberg, Nathan and Manuel Trajtenberg, "General Purpose Technology at Work: The Corliss Steam Engine in the Late 19th Century, *The Journal of Economic History*, vol. 64, no.1, (March 2004):

Ryan, Kathleen and Joy Chavez, "Media Report to Women," *Electronic News*, vol. 4, no.2, (2010).

Saporito, Bill, "Hack Attack," *Time*, vol. 178, no. 1, (July 4/2011).

Schneider, Davis, "Dial L for Location," *American Scientist*, vol. 93, no. 6, (November-December, 2005).

Schramm, Jeff, "Dropping the Fire: The Decline and Fall of the Steam Locomotive," *Technology and Culture*, vol. 42, no. 2, (April 2001).

Seccombe, Wally, "Patriarchy Stabilized: The Construction of the Male Breadwinner Wage Norms in the 19th Century," *Social History*, vol. 11, no. 1 (January 1986).

Sharpe ,William and Leonard Wallock, "Bold New City or Built-up Burb? Redefining Contemporary Suburbia," *American Quarterly*, Vol.46, No.1, (March 1994).

Sher, Lauren, "Katie Couric Returns to ABC," *Internet Ventures*, (June 6, 2011).

Shine, Edward, "The Junk Gun Predicament: Answers Do Exist," *Arizona State Law Journal*, vol. 30, (1998).

Shoup, G. Stanley, "The Control of International Radio Communication," *Annals of the American Academy of Political and Social Sciences*, vol. 142 (March 1929).

Singer, Ben, "Female Power in the Serial Queen Melodrama," *Camera Obscura*, vol. 22, (1990).

Steiger, William A., "Lindbergh Flies Air Mail from Springfield," *Journal of the Illinois State Historical Society*, vol. 47, no. 2, (Summer 1954).

Stern Charles, "PBS Tries to Keep Egg in Nest," Variety, (June 1998).

Strayer, David L., " A Comparison of the Cell phone Driver and the Drunk Driver, *Human Factors*, vol. 48, (2006).

Sumner, Daniel A., "Targeting Farm Programs," *Contemporary Policy Issues*, Vol 9, (January 1991).

Sweeney, Stephen B., "Some Economic Aspects of Aircraft Transportation," *The Annals of the American Academy*, vol. 131, (May 1927).

Thompson, Michael F., "Employment and Growth in the U.S. Automotive Manufacturing Industry," *Indiana Business Review*, vol. 85, no.1, (Spring 2010).

Townsend, Anthony M., "Life in the Real-time City: Mobile Telephones and Urban Metabolism," *Journal of Urban Technology*, vol. 7.

Tremblay, Richard E., "The Development of Aggressive Behavior During Childhood," *International Journal of Behavioral Development*, vol. 24, (2000).

Tulchinsky, Gerald, "Said to be a Very Honest Jew," *Urban History Review*, vol.18, no.3, (February 1990).

Turner, Joel, "The Messenger Overwhelming the Message," *Political Behavior*, vol. 29, no.4, (December 2007).

Urrey, John, "The 'System' of Automobility," *Theory, Culture and Society*, Vol.21, no.4-5, (2004).

Urry, John, "Mobility and Proximity," *Sociology*, vol.36 (2002).

Van West, Jeff, "It's About Engagement, " *IFR*, Vol. 27, no. 11, (November 2011).

Walters, William D., "The Geography of the European Dreadnaught Race," *Geographic Bulletin*, vol. 34, no.1, (1992).

Warren, Donald I., "Radio Priest: The Father of Hate Radio," *Michigan Sociological Review*, vol. 11, (Fall 1997).

Weiland, Hyman, "The Adolescent and the Automobile," *Chicago Review*, vol. 9, no..3, (Fall, 1955).

White, John H., "Septimus Norris and the Origin of the Ten Wheel Locomotive," *Technology and Culture*, vol. 9, no.1, (January 1968).

White, John H., "Industrial Locomotives: The Forgotten Servant," *Technology and Culture*, vol. 21., no. 2, (April 1980).

Williams, Rex, "The Porno Movie Scene," *Sensuous One*, (1973).

Wilson, Barbara, "Violence in Children's Television Programming," Journal of Communication, vol. 52, (2002).

Wilson, Melanie and Anita, "Gender and Teleworking Identities in the Risk Society," Work and Employment, vol.19, no. 3, (2004).

Wintermute, Garen J., "Relationship Between Illegal Use of Handguns and Handgun Sales Volume," *Journal of the American Medical Association*, vol. 284, no.5, (2000).

Wise, T. A., "Western Union by Grace of FCC and A.T. and T.," *Fortune*, (March 1979).

Witzel, H. D. and B.F. Vogelaar, "Engineering the Hillside Combine," *Agricultural Engineering*, vol. 36, (August 1955).

Woolheather, Linda, "The Advantage of Police Shooting Training Simulator," *Money*, (October 22, 2010).

Wortman, Susan, "The Unhealthy Business of Making Clothes," *Healthsharing*, vol 1, no11, (November 1979).

Yar, Majid, "The Novelty of Cybercrime," European *Journal of Criminology*, vol.2, (2005).

Yu, Gang, et. al. "Optimizing Pilot Planning and Training for Continental Airlines," *Interfaces*, vol. 34, no. 4, (July-August 2004).

Zehr, Mary Ann, "Laptops for All Doesn't Mean They are Always Used," *Education Week*, vol. 19, no,.30, (2000).

Zeitlin, Marilyn, "Too Old for Hollywood," *The Progressive*, vol. 50, no.1, (January 1992).

Zimmerman, Frederick, Dimitri Christakis and A.N. Meltzoff, "Television and DVD/Video Viewing in Children Younger than Two Years," *Archives of Pediatric and Adolescent Medicine*, vol. 161, no.5, (2007).

Zimmerman, Frederick J., Gwen M. Glew, Dimitri A. Christakis, and Wayne Katon, "Early Cognitive Stimulation, Emotional Support, and Television Watching," *Archives of Pediatric and Adolescent Medicine*, vol.159, (2005).

Zimring, Franklin E., "The Medium is the Message: Firearm Caliber as a Determinant of Death from Assault," *Journal of Legal Studies*, vol.1, no. 1. (1972).

Zinni, Clark, "Retrofitting Suburbia," *Alternatives Journal*, vol.26, no. 3 (2000).

Newspapers

Anderson, Dale and Phil Fairbanks, "49 killed as Plane Crashed into Home in Clarence Center," *The Buffalo News*, (February 13, 2009).

Barnard, Neal, Do We Eat Too Much Meat?" *The Huffington Post*, (January 12, 2011).

Barnes, Brooks, "A Year of Disappointment at the Movie Box Office," *The New York Times*, (December 25, 2011).

Bradley, Mark, "Out of Control: Behavior of Football Fans Takes and Ugly Turn," *Atlanta Journal Constitution*, (November 6, 2002).

Brody, Jane, "Teenage Risks and How to Avoid Them," The New York Times, (December 18, 2007).

Celizic, Mike, "Mom Tasered in Front of Kids Posed No Risk," *Today*, (August 14, 2009).

Chen, Brian X. and Nick Wingfield, "Apple Introduces Tools to Supplant Print," *The New York Times*, (January 20, 2012).

Clapper, Raymond, "Air Mail Action Viewed As End of Subsidies for Favored Few," *Washington Post*, (February 11, 1934).

Cody, Edward, "Pan Am Plane Crashes in Scotland, Killing 270," Washington Post, (December 22, 1988).

Cummings, Judith, "Neighbors Term Mass Slayer A Quiet but Hotheaded Loner," The New York Times, (July 20, 1984).

Daily Mail Reporter, "No Hiding Place from New U.S. Army Rifles that Use Radio Controlled Smart Bullets," Mail Online, (November 30, 2010).

de Moraes, Lisa, "Super Bowl XLVI: Biggest TV audience ever," *The Washington Post*, (February 6, 2012).

Dicks, Nikasha, "Families Join GasCar to Share Love of Automobiles," *The Augusta Chronicle*, (October 8, 2009).

Dillon, Sam, "Many Nations Passing U.S. in Education, Expert Says," *The New York Times*, (March 10, 2010).

Dobbs, Michael, "IBM Technology Aided the Holocaust," *The Washington Post*, (February 11, 2001).

Dougherty, Philip, "New Pitch to Spur Phone Use," *The New York Times*, Sec. D. October 23, 1985).

Editorial Staff, "Ignoring the Lawlessness,"*Augusta Chronicle*, (October 30, 2011)

Editorial, "Megrahi Inquisition," *Daily Mail*, (December 10, 2010).

Editorial, "The Fire That Changed Everything," *The New York Times*, February 23, 2011.

Engstrom, Tim, "A Look at International Harvester History," *The Austin Daily Herald*, (May 17, 2011).

Gertz, Bill, "Inside the Ring: PLA Hack on Google?" *Washington Times*, (July 8, 2010).

Gorman, Siobhan and Julian E. Barnes, "Cyber Combat: Act of War," *The Wall Street Journal*, (May 30, 2011).

Hacker, Andrew, "The Hollywood Influence," *The New York Times*, (November 9, 1973).

Hannah, Jim, "Student Dies After Police use Taser at University of Cincinnati," *The Cincinnati Enquirer*, August 7, 2011).

Hernandez, Raymond, "What Would you Drive if the Taxpayer Paid?" *The New York Times*, (May 1, 2008).

Hodges, R.H., "Facts and Policy on Police Guns," *The New York Times*, (April 2, 1989).

Holusha, John, "Mass Killer is Recalled as a Gun-Loving Youth," *The New York Times*, (July 21, 1984).

Hunter, Stephen, "Triggering Memories: At the NRA Museum, Tommy Gun Devotees Can Zero In on a Classic," The *Washington Post*, (March 22, 2004).

Jalonick, Mary Clare, "Farm Subsidies: House Appropriations Committee Approves Cuts," *The Huffington Post*, (May 31, 2011).

Jordan, Jim, "Radio's Fibber McGee," *The New York Times*, (April 8, 1961).

Keh, Andrew, "Dugout Phones Last Bastion of the Landlines," *The New York Times*, (October 22, 2011).

Keyes, David, "Industrial Revolution was Powered by Child Slaves," *The Independent*, (August 2, 2010).

King, Mackenzie, "Foremen demanding bribes," *The Daily Mail*, (Toronto: October 9th, 1897).

Kovaleski, Serge F., "Havana is Haven for Fugitive '70s Hijacker," *Washington Post*, (August 31, 1999).

Kritzer, Jamie, "Are you Ready for some Football?" *Greenboro News and Record*, (September 12, 2002).

Kunkle, Frederick, "Gun Enthusiasts Find a Paradise of Wood and Steel," *The Washington Post*, (June 25, 2006).

Lemlich, Lara, "Life in the Shop," *New York Evening Journal*, (November 28, 1909).

Lesmerises, Doug, "In Sports, NFL Means Business," *The Wilmington News Journal*, (January 27, 2003).

Lomatire, Paul, "Have I Got News for You About Molly," *The Palm Beach Post*, (June 18, 1994).

Maslin, Janet, "The Secret Lives of Dangerous Hackers," *The New York Times*, (June 1, 2012).

Matsumoto, Rick, "Stock Car Racing Well on Way to Outgrowing Redneck Roots," *The Toronto Star*, (April 5, 1997).

McCartin, Joseph, "Ronald Reagan, the Air Traffic Controllers Association and the Strike that Changed America," *The New York Times*, (August 3, 2011).

Meckel, Miriam, "Mensch wird Maschine," *Die Zeit*, (June 12, 2012).

Mouawad, Jad, "Trapped in the Middle Seat," *The New York Times*, (May 3, 2012).

Negroni, Christine, "Business Jet Owners Say Proposed Rules Won't Improve Security," *The New York Times*, (January 11, 2009).

Negroni, Christine, "What to Expect as Airport Security Rules are Tightened," *The New York Times*, (January 2, 2010).

No author, "A Day in the Life of America," *The New York Post*, (September 12, 2001).

No author, "Army Ill-equipped For Mail Service, Officer Predicted," *San Diego Union*, (March 18, 1934).

No author, "Death by Police Taser," *San Jose Mercury News*, (February 2, 2009).

No author, "Farmers Telephone at Low Cost," *Chicago Daily Tribune*, (September 20, 1903).

No author, "Garment Manufacturers meeting in New York City," *Minnesota Union Advocate*, (January 21, 1925).

No author, "Hijackers To Cuba Given Stiff Terms," *The New York Times*, (July 7, 1983).

No author, "LI Man dies after Tasering," *New York Post*, (May 7, 2001).

No author, "Roxxxy Sex Robot World's First Robot Girlfriend Can Do More Than Chat," *Huffington Post*, IM arch 18, 2006).

No author, "The Story of the Sewing Machine," *The New York Times*, (January 7, 1860).

Reagan, Gillian, "Stitch by Stitch and Block by Block," *The New York Times*, (February 17, 2011).

Richter, Matt, "Drivers and Legislators Dismiss Cellphone Risks," *New York Times*, (July 19, 2009).

Richter, Matt, "In Study, Texting Lifts Crash Risk by Large Margin," *New York Times*, (July 28, 2009).

Roberts ,Sam, "Population Study Finds Change in the Suburbs," *The New York Times*, (May 9, 2010).

Russakoff, Dale, "An Ordinary Boy's Extraordinary Rage," *The Washington Post*, (July 2, 1995).

Sharkey, Jo, "Private Jets Warming Up for Presidential Campaign," *The New York Times*, (April 5, 2011).

Sharkey, Jo, "When a Pat-Down Seems Like Groping," *The New York Times*, (November 2, 2004).

Sheng, Ellen, "Sirius Bolsters Lineup to Battle FM," *Wall Street Journal*, (June 2, 2004).

Simon, Greg, "Sunday Dialog: Using Technology to Teach," *The New York Times*, (October 20, 2011).

Stein, Robert and Donna St. George, "Unwed Motherhood Increases Sharply in U.S." *The Washington Post*, (May 14, 2009).

Stock, Ulrich, "Das Reis-Phone," *Die Zeit*, (October 15, 2011).

Thomma, Steven, "Study: Polls that Leave Out Cell Phones Skew Results Toward GOP" *McClatchy Newspapers*, (October 18, 2010).

Trebay, Guy, "Needle and Thread Still Have a Home," (The New York Times, April 28, 2010).

Weigel, Jen, "Cybercrime: A Billion Dollar Industry," *Chicago Tribune*, (September 26, 2011).

Weiner, Tim, "Manzanillo Journal: In Corn's Cradle, U.S. Imports Bury Family Farms," *New York Times*, (February 26, 2002).

Weingarten, Marc, "To Stitch, Take Needle, Thread and a Computer: New Machines Revolutionize Sewing," *The New York Times*, (August 27, 2000).

Welter, Greg, "First Sky Jacking Attempt Was In Chico, 45 Years Ago," *Chicoer*, (July 31, 2006).

Wysocki, Beata, "Many Youths Splurge Mostly on US Goods," *Wall Street Journal*, (June 26, 1997).

Yusko, Dennis, "Taser used in Spa City detailed," *Times Union*, (Albany, N.Y.) July 14, 2011.

Documents

"Aircraft Production in the United States, S. Rept. 555, Congressional Record, 65th Congress, 2nd Session,9239-34.

American Council of Education v. FCC, 451 F. 3rd 226, 38 Communications Reg. (P&F) 859.

Annual Report of the Bureau of Industry, (Toronto: 1889).

Anthropometric Reference Data for Children and Adults: United States, 2003-2006, Centers for Disease Control and Prevention, National Center for Health Statistics, (March 6, 2009).

Bachu, Amara, "Trends in Marital Status of U.S. Women at First Birth: 1930-1994" U.S. Bureau of the Census, Population Division, Working Paper No. 20 (March 1998).

Bowes, Douglas E., Wayne D. Rasmussen, and Gladys L. Baker, "History of Agricultural Price Support and Adjustment Programs, 1933-1948," (Washington, DC The U.S. Department of Agriculture, Information Bulletin 485,1985),.

Bureau of Labor Statistics, "Occupational Employment and Wages," (Washington, DC, United States Government Printing Office, May 2009).

Bureau of Labor Statistics, U.S. Department of Labor, Occupational Outlook Handbook (Washington D.C. The United States Government Printing Office, 2010-2011 Edition, "Earnings."

Center for Disease Control, "Key Birth Statistics" (Atlanta, Ga. April 6, 2011).

Christain v. Johns, 125 Ga. 977 1904 and Smith v. Jordan, 211 Mass. 269 (1912).

Constitution of the United States of America, Fourth Amendment.

Department of Health and Human Services, National Center for Health Statistics, National Vital Statistics Reports, vol. 54., no.19, (June 28, 2006). See also: U.S. Census Bureau, 2008 National Population Projections (August 2008).

Federal Communications Commission, "Implementation of 911 Act," Washington D.C. 15 F.C.C.R17079, paragraph 1, 20 Com. Reg. 489 (2000).

Gardner, Bruce L., "Does the Economic Situation of U.S. Agriculture Justify Commodity Support Programs?" AEI Agricultural Policy Series: The 2007 Farm Bill and Beyond. (Washington D.C. The American Enterprise Institute, 2007).

Hing, Jennifer, "Appropriations Committee Approves Fiscal Year 2012 Agriculture Bill," *The Capitol*, Washington D.C. "H 307 http://appropriations.house.gov/news/ documentprint.aspx?DocumentID=244890 (May 31, 2011).

Hollender, Jay Michael, "Prelude to a Strike," Proceedings of the New Jersey Historical Society, vol. 79, (July 1961).

Interstate Circuit v. City of Dallas, 249 F. Supp. F. Supp. 19 (N.D. Tex. 1965).

Johnson, Renee and Jim Monke, "What Is the Farm Bill?" Congressional Research Service Report for Congress, (December 10, 2010).

Knowles, Paul F., "Flax Production in the Imperial Valley," California Agricultural Experiment Station Circular, Davis, California No.480, (July 1959).

Levittown Historical Society, A Brief History of Levittown, New York, (New York: Levittown Historical Society, No date).

Lubowski, Ruben N., Marlow Vesterby, Shawn Bucholtz, Alba Baez, and Michael J. Roberts, "Major Uses of Land in the United States," United States Department of Agriculture, Economic Research Service, (2002).

McCormick Collection, McCormick Reaper Centennial Source Material, (Chicago: State Historical Society of Wisconsin, 1931).

Minnesota Bureau of Labor Statistics, Eighth Biannual Report, 1901-1902.

Mowchan, John A., "Don't Draw the Red Line," U.S. Naval Institute Proceedings, vol. 137, no. 10, (October 2011).

National Highway Traffic Safety Administration, "Choose Responsibility," Washington, D.C. (2010).

Newport, Frank, "America's Church Attendance Inches Up in 2010," (Princeton, NJ: Gallup Poll, http://www.gallup.com/poll/141044/americans-church-attendance-inches-2010.aspx, June 25, 2010).

No author, "Cosmetic Plastic Surgery Research," New York, American Society for Aesthetic Plastic Surgery, ttp://www.cosmeticplasticsurgerystatistics.com/statistics.html (2011).

No author, "Deaths in World Trade Center Terrorist Attacks," Morbidity and Mortality Weekly Report, (Atlanta, GA: Centers for Disease Control and Prevention, September 11, 2002)..

No author, "Fatalities and Fatal Rates by Quarter," Traffic Safety Facts, (Washington, DC, The United States Printing Office, (December 2010).

No author, "Team Partnershuip: Improving Steam System Efficeincy," U.S. Department of Energy, (2000).

No suthor, National Highway Safety Administration, U.S. Department of Transportation, "Using Wireless Communication Devices While Driving," (2003).

No author, United States Department of Agriculture, "Ágricultural Exports and Net Income 1997-2005" USDA economic Research Service, (February 2006).

No author, "What is Genetic Engineering?" Union of Concerned Scientists, (2003).

Pittman, Harrison M., "The Constitutionality of Corporate Farming Law in the Eighth Circuit," The National Agricultural Law Center, (June 2004).

Public Law 75-718, ch. 676, 53 Stat 1060, June 25, 1938, 29 USC.

Radio Act of 1927, Public Law no. 632, Chapter 169, 44 Stat. 1172.

Roth v. U.S., 354 U.S. 476 (1957)

Second Amendment to the Constitution of the United States as ratified December 15, 1791.

Spielmaker, Debra, A History of American Agriculture, (Proceedings of the AAAE Research Conference, vol. 34. Minneapolis, MN, 2007)

Sueper, Paul, "Number of Farms and Farms by Sales Class," Washington, D.C. United States Department of Agriculture, National Agriculture Statistics Service, (February 18, 2011).

Tarmann, Allison, "Fifty Years of Demographic Change in Rural America," Washington, D.C. Population Reference Bureau, 2009).

Tavernetti, James R. and Lyle M. Carter, "Mechanization of Cotton Production," (Davis, CA: California Agricultural Experiment Station, Bulletin 804, 1983).

The Statutes At Large of the United States of America, vol. XLIV, Part 2, Sixty-ninth Congress, 1925-1927 Session I.

U.S. Bureau of Labor Statistics, (Washington DC: Occupational Employment and Wages, 2012).

U.S. Bureau of the Census, Historical Statistics of the United States, (Washington, DC: U.S. Government Printing Office, 1975).

U.S. Bureau of the Census, Population Division, "Immigration Volume and Rates, (March 9, 1999).

U.S. Bureau of the Census, Statistical Abstract of the United States, (Washington, DC: U.S. Government Printing Office, 1984).

U.S. Census Bureau, (Washington DC: Statistical Abstracts of the United States, (2011)..

U.S. Congress, 73rd Congress, Session II, Chapter 757.

U.S. Department of Agriculture, Corporations with Farming Operations, Agricultural Economics Report No. 209, (Washington, DC: U.S. Government Printing Office, June 1971).

U.S. Department of Agriculture, Economic Research Service and Foreign Agricultural Research Service "Foreign Agricultural Trade of the United States" (February 2010), Table 850. U.S. Census Bureau Statistical Abstract of the United States. (2011).

U.S. Department of Commerce, Bureau of the Census, "Expectation of Life at Birth," Statistical Abstracts, (2012).

U.S. Department of Commerce, Bureau of the Census, Fourteenth Census of the United States, vol. XI, Mines and Quarries, (1913).

U.S. Department of Commerce, U.S. Census Bureau, (Washington, DC: U.S. Census 2010).

U.S. Department of Justice, Federal Bureau of Investigation, Uniform Crime Report, (Washington, DC, 2009).

U.S. Department of Labor, "Usual Weekly Earnings of Wage and Salary Workers, Fourth Quarter, 2010," (Washington, DC: Bureau of Labor Statistics, 2011).

U.S. Department of Labor, Bureau of Labor Statistics, "Real Earnings September 2010" Table A-1.

U.S. Department of Transportation, Federal Aviation Administration, CFR Part 61,"Certification: Pilots, Flight Instructors and Ground Instructors," Sub-part A General Section 61.

U.S. Federal Highway Administration, "Highway Statistics 2008," TableHM 12.

U.S. House of Representatives, 52nd Congress, Second Session, Committee on Manufactures, "The Sweating System," vol. 1, no.1, (Washington, DC: The United States Government Printing Office,1893).

U.S. Patriot Act, (2001) Public Law no. 107-56115.Stat.

United States Department of Labor, Bureau of Labor Statistics, Minimum Quantity Budget Necessary to Maintain a Workers Family of Five at a Level of Health and Decency, (Washington, D.C. Government Printing Office, 1920).

United States Government, Environmental Protection Agency, "Ag 1010" Washington, DC, (Government Printing Office, 2007).

United States Government, Gun Control Act of 1968, 18 U.S.C. Chapter 44.

USDA Economic Research Service , "Americans Spend Less than 10% of Disposable Income on Food," (Juu 19, 2006).

Vilsack, Tom, "US Farm Exports Set Record in 2011," The Department of Agriculture-Economic Research Service, (Washington, DC 2011).

U.S. Bureau of Transportation, Bureau of Transportation, Research and Innovative Technology Administration, "Trans Stats," (2011).

Watkins v. Clark, 103 Kans. 629 (1918).

Weatherford, Brian, The State of U.S. Railroads, (RAND Corp., 2008).

Yung T.J. and B.K. Yamaoka, A National Directory of 911 Systems, Washington, D.C. Bureau of Justice Statistics, (1980).

Index

United Mine Workers of America,
    10
United States, 2, 20, 144
United States government, 188
United States Marines, 31
United States Naval Observatory,
    98
United States Navy, 89
United States Patent Office, 116
United States v Miller, 77
University of California, 42
University of Chicago, 150
University of Cincinnati, 88
University of Texas at Austin, 144
upper middle class status, 27
urban areas, 198
urbanization, 2
Van Nuys, California, 162
Veblen, Thorstein, 28
victimization, 66
video game, 179
video game players, 179
violence, 124, 148
violence on television, 92
Wal-Mart, 71
Walt Disney Pictures, 138
War of 1812, 8
War of the Worlds, 119
Washington and Baltimore, 98

Washington, DC, 197
Watt. James, 1
well ordered police state, 191
Western Europe, 26
Western Union, 98, 102
White, Andrew Dickson, 201
white collar jobs, 200
white knuckle, 166
White Label, 69
white population, 26
wildlife management, 85
William Paterson College, 64
Williamsburg seamstress, the, 71
Wilson., Woodrow, 99-101
Winfrey, Oprah, 121
wiretapping, 109
Women's Trade Union League, 67
work colleagues, 183
work from home, 185
work in the factories, 68
working class, 104
World Trade Center, 167
World Trade Organization, 49
World Wide Web, 177
Wright brothers, 156
X rated caricatures, 139
young women, 68
Zimmermann telegram, 99